Student Solutions Manual
to accompany

PRINCIPLES OF PHYSICS

SERWAY

Ralph V. McGrew
Broome Community College

Steve Van Wyk
Chapman College

Raymond A. Serway
James Madison University

Saunders College Publishing

SAUNDERS GOLDEN SUNBURST SERIES
Harcourt Brace College Publishers

Fort Worth   Philadelphia   San Diego   New York   Orlando   San Antonio
Austin   Toronto   Montreal   London   Sydney   Tokyo

McGrew/Van Wyk/Serway: Student Solutions Manual to
accompany Principles of Physics by Raymond A. Serway

ISBN 0-03-098865-9

456 095 987654321

Preface

Don't read this book passively. Use it actively to guide and check your own work, if you want extra practice solving problems in your physics course with the textbook **Principles of Physics** by Raymond A. Serway. Do you need extra practice solving problems? Talk with successful former students or with your instructor; along with a short answer they may help you see how valuable problem-solving skill is in your curriculum, in your field.

To study one of the chapters, I recommend you preview, carefully read, and review the textbook chapter; then attend a class; and then start doing problems. Do problems every day. Begin with the examples incorporated into the textbook chapter: cover up the solution and work each one through yourself. If an exercise immediately follows the example, do it on paper and immediately check your answer. Now proceed to the homework questions and problems your instructor assigns. Write, sketch, list, try out, cross out, start again the next day, work with other students, ask the teacher, don't quit, think -- for some problems you will need to buckle on the whole armor of truth. Finally, you can use this book to guide more practice. This manual contains answers for a few of the questions, and full solutions for several of the problems at the end of each chapter. Assign some of these to yourself as homework and again write out your own solutions, with this book firmly closed. If a problem is difficult, try it, think about it, and try it again the next day. At last open this book and compare your response with ours. Take notes in a different color. Gloat over how well you did, or figure out how to get over or around the parts that are tough for you.

Your solution to a problem does not have to look just like ours. A problem can sometimes be solved in different ways, starting from different principles. If you wonder about the validity of an alternative approach, ask your instructor.

We try to make each solution complete, starting from general principles, stating the physical reasoning, and showing mathematical steps. In calculations we assume the data is precise enough to permit three significant digits in answers. If there are several steps, we compute intermediate results to several digits, even though we write down only three. We "round off" only when we write down an answer at the end, never anywhere in the middle of a chain of steps.

We thank Sue Jarrett and Judy Sarver for computer typesetting this book. We thank Michael Rudmin for producing diagrams. We thank Jennifer Bortel, Senior Developmental Editor, for coordinating this project.

With the publisher, we try to produce an error-free book. If you find a mistake, please send a note to me:
Ralph McGrew
Engineering Science Department
Broome Community College
Box 1017
Binghamton, New York 13902-1017

Good luck in your challenging and rewarding work!

TABLE OF CONTENTS

Chapter 1 1

Chapter 2 10

Chapter 3 26

Chapter 4 41

Chapter 5 56

Chapter 6 74

Chapter 7 83

Chapter 8 94

Chapter 9 110

Chapter 10 126

Chapter 11 137

Chapter 12 156

Chapter 13 167

Chapter 14 179

Chapter 15 192

Chapter 16 208

Chapter 17 224

Chapter 18 241

Chapter 19 259

Chapter 20 275

Chapter 21 290

Chapter 22 302

Chapter 23 315

Chapter 24 325

Chapter 25 336

Chapter 26 351

Chapter 27 365

Chapter 28 377

Chapter 29 388

Chapter 30 400

Chapter 31 413

Chapter 32 425

CHAPTER 1

QUESTIONS

14   **QUESTION:** The magnitudes of two vectors **A** and **B** are $A = 5$ units and $B = 2$ units. Find the largest smallest values possible for the resultant vector **R** = **A** + **B**.

ANSWER: If **A** and **B** are in the same direction, then $|A + B| = R = 7$ units. If **A** and **B** are in opposite directions, then $|A + B| = R = 3$ units

27   **QUESTION:** While traveling along a straight interstate highway you notice that the mile marker reads 260. You travel until you reach the 150-mile marker and then retrace your path to the 175-mile marker. What is the magnitude of your resultant displacement from the 260-mile marker?

ANSWER: Your first displacement is 110 miles in the direction of decreasing numbers. Your second displacement is 25 miles in the opposite direction. So your total displacement is 85 miles in the direction of decreasing numbers, with magnitude 85 miles □.

29   **QUESTION:** A roller coaster travels 135 ft at an angle of $40°$ above the horizontal. How far does it move horizontally and vertically?

ANSWER: Its horizontal displacement component is
$(135 \text{ ft}) \cos 40° = 103$ ft □
It attains a height $(135 \text{ ft}) \sin 40° = 86.8$ ft □

CHAPTER 1

PROBLEMS

1.1   **PROBLEM:** Calculate the density of a solid cube that measures
5 cm on each side and has a mass of 350 g.

SOLUTION: The volume of the cube is $(5 \text{ cm})^3 = 125 \text{ cm}^3$. So its
density is $\rho = m/V = 350 \text{ g}/125 \text{ cm}^3 = 2.80 \text{ g/cm}^3$. To convert
this to standard units of kg/m^3, we note that 10^3 g = 1 kg, and
ONE MILLION cubic centimeters make one cubic meter:
$1 \text{ m}^3 = (100 \text{ cm})^3 = 10^6 \text{ cm}^3$.

So, $\rho = 2.80 \dfrac{g}{cm^3}\left(\dfrac{1 \text{ kg}}{10^3 \text{ g}}\right)\left(\dfrac{10^6 \text{ cm}^3}{1 \text{ m}^3}\right) = 2.80 \times 10^3 \text{ kg/m}^3$ $\square$

1.11   **PROBLEM:** Newton's law of universal gravitation is given by

$$F = G \frac{Mm}{r^2}$$

Here F is the force of gravity, M and m are masses, and r is a
length. Force has the units $kg{\cdot}m/s^2$. What are the SI units of
the proportionality constant G?

SOLUTION: The units in the equation $F = GMm/r^2$ require
$kg{\cdot}m/s^2 = [G] \text{ kg}{\cdot}kg/m^2$ where $[G]$ represents the units of G.
Solving, we have $[G] = m^3/(s^2{\cdot}kg)$ $\square$

1.17 **PROBLEM:** (a) Find a conversion factor to convert from miles per hour to kilometers per hour. (b) Until recently, federal law mandated that highway speeds would be 55 mi/h. Use the conversion factor of part (a) to find the speed in kilometers per hour. (c) The maximum highway speed has been raised to 65 mi/h in some places. In kilometers per hour, how much increase is this over the 55-mi/h limit?

SOLUTION: You can look up relationships between units in Appendix A, but it is easier to use the facts you might remember:

1 mile = 5280 ft, 1 ft = 12 in, 1 in = 2.54 cm, $c = 10^{-2}$, $k = 10^3$.

Then,

$$1\,\frac{mi}{h} = \frac{mi}{h}\left(\frac{5280\ ft}{1\ mi}\right)\left(\frac{12\ in}{1\ ft}\right)\left(\frac{2.54\ cm}{1\ in}\right)\left(\frac{1\ m}{10^2\ cm}\right)\left(\frac{1\ km}{10^3\ m}\right)$$

$$= 1.61\ km/h$$

(b) $\quad 5\,\dfrac{mi}{h} = 55\,\dfrac{mi}{h}\ \dfrac{1.61\ km\,/\,h}{1\ mi\,/\,h} = 88.5\ km/h$ □

(c) The increase is (65 mi/h) - (55 mi/h) = 10 mi/h

$$\Delta v = 10\,\frac{mi}{h}\left(\frac{1.61\ km\,/\,h}{1\ mi\,/\,h}\right) = 16.1\ km/h\ □$$

1.25   **PROBLEM:**   One cubic meter (1.0 m^3) of aluminum has a mass of 2.70 x 10^3 kg, and 1.0 m^3 of iron has a mass of 7.86 x 10^3 kg. Find the radius of a solid aluminum sphere that will balance a solid iron sphere of radius 2.0 cm on an equal-arm balance.

SOLUTION: The statement of the problem tells us the densities of the materials ρ_{Al} = 2.70 x 10^3 kg/m^3 and ρ_{Fe} = 7.86 x 10^3 kg/m^3.   To balance, the spheres must have equal masses: m_{Al} = m_{Fe}.   The definition of density $\rho = m/V$ then gives $\rho_{Al}V_{Al} = \rho_{Fe}V_{Fe}$.

And the volume of a sphere is   4/3)πr^3

So at last, $\rho_{Al}(4/3)\pi r_{Al}^3$ = $\rho_{Fe}(4/3)\pi r_F^3$ and

$r_{Al} = r_{Fe}(\rho_{Fe}/\rho_{Al})^{1/3}$

r_{Al} = (2 cm)(7.86 x 10^3 kg/m^3/2.70 x 10^3 kg/m^3)$^{1/3}$

   = (2 cm)(7.86/2.70)$^{1/3}$ = 2.86 cm □

An alternative method of solving the problem would be to find numerically the volume of the iron sphere (33.5 cm^3), it mass (0.263 kg), and the volume of the aluminum sphere (97.6 cm^3), to work back to its radius.   But get used to the method we presented first.   It is better because it is numerically simpler.

1.27   **PROBLEM:**   Estimate the amount of motor oil used by all cars in the United states each year and its cost to the consumer.

SOLUTION: With a population around 250 million, the nation has ˜ 10^8 cars; that is, closer to 10^8 than to 10^7 or to 10^9.   If each uses ˜ 10^1 quarts of motor oil per year, the nation must refine and dispose of ˜ 10^9 quarts per year □. Each quart costs closer to $1 than to 10¢ or $10, so the cost is   ˜ $10^9/y □.

1.35 **PROBLEM:** How many significant figures are there in

(a) 7.89 ± 0.2, (b) 3.788 x 10^9, (c) 2.46 x 10^{-6}, (d) 0.0053?

SOLUTION: The significant digits are underlined:

(a) <u>78.9</u> ± 0.2 has 3.

(b) <u>3.788</u> x 10^9 has 4.

(c) <u>2.46</u> x 10^{-6} has 3.

(d) 0.00<u>53</u> has 2.

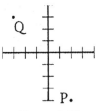

1.37 **PROBLEM:** Two points in the *xy* plane have cartesian coordinates (2.0 - 4.0) and (-3.0, 3.0), where the units are meters. Determine (a) the distance between these points and (b) their polar coordinates.

Figure 1.37

SOLUTION: The position vector for the first point is the vector from the origin to it: $\mathbf{r}_P = (2\mathbf{i} - 4\mathbf{j})$m. The position vector of the second point is $\mathbf{r}_Q = (-3\mathbf{i} + 3\mathbf{j})$m. The displacement <u>from</u> the first point <u>to</u> the second is
$$\mathbf{r}_Q - \mathbf{r}_P = (-3\mathbf{i} + 3\mathbf{j})\text{m} - (2\mathbf{i} - 4\mathbf{j})\text{m} = (-5\mathbf{i} + 7\mathbf{j})\text{m}$$

so its magnitude is the distance

$$\sqrt{(-5 \text{ m})^2 + (7 \text{ m})^2} \quad = 8.60 \text{ m } \square$$

(b) $\mathbf{r}_P = (2\mathbf{i} - 4\mathbf{j})\text{m} = \sqrt{(2 \text{ m})^2 + (-4 \text{ m})^2}$ at Arctan (-4 m)/(2 m)

$\mathbf{r}_P = 4.47$ m at -63.4° $\square$

$\mathbf{r}_Q = (-3\mathbf{i} + 3\mathbf{j})\text{m} = \sqrt{(-3 \text{ m})^2 + (3 \text{ m})^2}$ at Arctan (3 m)/(3 m)
above the -*x* axis

$\mathbf{r}_Q = 4.24$ m at 45.0° above the -*x* axis

$\mathbf{r}_Q = 4.24$ m at 135° $\square$

Note that on most calculators, if you try to find the direction of $\mathbf{r}_Q$ as Arctan (3m/-3m), you will get the WRONG answer -45° = 315°. You need to think about how its angle fits in with the coordinate axes.

1.45 **PROBLEM:** A roller coaster moves 200 ft horizontally, then rises 135 ft at an angle of 30° above the horizontal. It then travels 135 ft at an angle of 40° downward. At the end of this movement, what is its displacement from its starting point? Use graphical techniques.

SOLUTION: A convenient scale is 1 cm = 50 ft. Lay out the path of the car, stringing the vectors along tail-to-head, with protractor and ruler:

This gives **R** = 4.2 x 10² ft at -3° □.
With graphical techniques we claim no better precision.

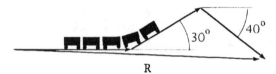

$$30° \qquad 40°$$

R

Figure 1.45

1.55 **PROBLEM:** A jet airliner moving initially at 300 mph due east enters a region where the wind is blowing at 100 mph in a direction 30° north of east. What are the new velocity and direction of the aircraft?

SOLUTION: The wind blows the plane along, making its total velocity the vector sum of the velocity of air relative to ground and the velocity of plane relative to air. Take the x-axis east and y-axis north: (300 mi/h)**i** + (100 mi/h) at 30°

v = (300 mi/h)**i** + (100 mi/h) cos 30° **i** + (100 mi/h) sin 30° **j**

    = (300 mi/h)**i** + (86.6 mi/h)**i** + (50.0 mi/h)**j**

    = (387 mi/h)**i** + (50.0 mi/h)**j**

This is the velocity vector. To find its direction, we proceed:

    v = $\sqrt{387^2 + 50^2}$ mi/h at Arctan 50/387

    v = 390 mi/h at 7.37° N of E □

1.59 **PROBLEM:** Three vectors are oriented as shown in Figure 1.59, where $|\mathbf{A}| = 20$, $|\mathbf{B}| = 40$, and $|\mathbf{C}| = 30$ units. Find (a) the x and y components of the resultant vector and (b) the magnitude and direction of the resultant vector.

SOLUTION:

$\mathbf{A} + \mathbf{B} + \mathbf{C} = 20\mathbf{j} + (40 \text{ at } 45°) + (30 \text{ at } -45°)$

$\quad = 20\mathbf{j} + 40\cos(45°)\mathbf{i} + 40\sin(45°)\mathbf{j}$

$\qquad + 30\cos(-45°)\mathbf{i} + 30\sin(-45°)\mathbf{j}$

$\quad = 20\mathbf{j} + 28.3\mathbf{i} + 28.3\mathbf{j} + 21.2\mathbf{i} - 21.2\mathbf{j}$

[Now is a good time to check the signs you have written down against the arrowheads in the figure.]

$\mathbf{A} + \mathbf{B} + \mathbf{C} = 49.5\mathbf{i} + 27.1\mathbf{j}$ □

(b) $\mathbf{A} + \mathbf{B} + \mathbf{C} = \sqrt{49.5^2 + 27.1^2}$ at Arctan 27.1/49.5

$\qquad\qquad = 56.4 \text{ at } 28.7°$ □

1.61 **PROBLEM:** A useful fact is that there are about π x 10^7 s in one year. Use a calculator to find the percentage error in this approximation.

Note: percent error = $\dfrac{|\text{assumed value} \;-\; \text{true value}|}{\text{true value}}$ x 100

SOLUTION: The number of seconds in one year is

$$1 \text{ y} = 1 \text{ y}\left(\frac{365.25 \text{ d}}{1 \text{ y}}\right)\left(\frac{24 \text{ h}}{1 \text{ d}}\right)\left(\frac{3600 \text{ s}}{1 \text{ h}}\right) = 3.1558 \text{ x } 10^7 \text{ s}$$

so the error is

$$\left|\frac{3.14159 \text{ x } 10^7 \text{ s } - \; 3.1558 \text{ x } 10^7 \text{ s}}{3.1558 \text{ x } 10^7 \text{ s}}\right| = \frac{3.1558 - \pi}{3.16}$$

$$= 4.49 \text{ x } 10^{-3} = 4.49 \text{ x } 10^{-3} \text{ x } 100\% = 0.449\% \;\; \square$$

1.67 **PROBLEM:** A person going for a walk follows the path shown in Figure 1.23. The total trip consists of four straight-line paths. At the end of the walk, what is the person's resultant displacement, measured from the starting point?

SOLUTION:

$\mathbf{R}$ = (100 m)$\mathbf{i}$ + (300 m)(-$\mathbf{j}$) + 150 m at 210° + 200 m at 120°

$\quad$ = (100 m)$\mathbf{i}$ - (300 m)$\mathbf{j}$ - (150 m)cos 30° $\mathbf{i}$

$\qquad$ - (150 m)sin 30° $\mathbf{j}$ - (200 m)cos 60° $\mathbf{i}$

$\qquad$ + (200 m)sin 60° $\mathbf{j}$

$\quad$ = (100 m)$\mathbf{i}$ - (300 m)$\mathbf{j}$ - (130 m)$\mathbf{i}$ - (75.0 m)$\mathbf{j}$ - (100 m)$\mathbf{i}$

$\qquad$ + (173 m)$\mathbf{j}$

$\quad$ = (-130 m)$\mathbf{i}$ - (202 m)$\mathbf{j}$

This is the answer for total displacement. If you botched anything, you botched the signs, didn't you? They are not hard to get right if you PAY ATTENTION TO THE PICTURE. The displacement can also be expressed in magnitude-and-direction form

$$\mathbf{R} = \sqrt{(-130 \text{ m})^2 + (-202 \text{ m})^2}$$

$$\text{at Arctan } \frac{-202 \text{ m}}{-130 \text{ m}} \text{ below the } -x \text{ axis}$$

$\mathbf{R}$ = 240 m at 57.2° below the -x axis

$\mathbf{R}$ = 240 m at 237° $\square$

CHAPTER 2

QUESTIONS

7   **QUESTION:** Can the equations of kinematics (Eqs. 2.7 through
      2.11) be used in a situation where the acceleration varies in
      time? Can they be used when the acceleration is zero?

      ANSWER: If the acceleration changes in steps, having one
      constant value for a while and then another constant value for
      some time interval afterwards, we can use the equations for
      constant-acceleration motion to follow each section of the motion
      separately. If the acceleration is zero for a time interval, the
      four equations do apply to this constant-velocity motion,
      simplifying to $v = v_0$ and $x - x_0 = vt$. If the acceleration
      varies continuously, then the equations of kinematics cannot be
      used.

16   **QUESTION:** A ball is thrown upward. While the ball is in the
      air, (a) does its acceleration increase, decrease, or remain
      constant? (b) Describe what happens to its velocity.

      ANSWER: (a) The acceleration is constant in magnitude and
      direction throughout the free flight of the ball, from the
      instant after it leaves your hand until the instant before it
      strikes the ground. It is the downward acceleration of gravity.
      (b) The velocity changes continuously, having 9.8 m/s subtracted
      from it during each second of motion. So the upward speed of the
      object decreases to zero; then the object moves down with
      increasing-magnitude negative velocity.

20   **QUESTION:**   A ball rolls in a straight line along the horizontal direction. Using motion diagrams (or multiflash photographs) as in Figure 2.10, describe the velocity and acceleration of the ball for each of the following situations: (a) The ball moves to the right at a constant speed.   (b) The ball moves from right to left and continually slows down.   (c) The ball moves from right to left and continually speeds up.   (d) The ball moves to the right, first speeding up at a constant rate, and then slowing down at a constant rate.

ANSWER:

We use ⟶ for velocity, and ⟹ for acceleration. Read the stop-action pictures from left to right.

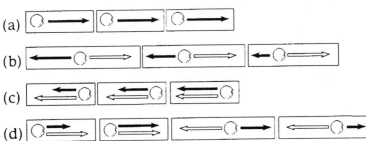

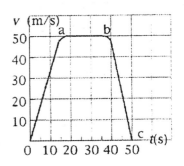

Figure 2.20

22   **QUESTION:**   A pebble is dropped into a water well, and the splash is heard 16 s later, as illustrated in the cartoon strip on page 57.   What is the *approximate* distance from the rim of the well to the water's surface?

ANSWER: Ignore air resistance, in 16 s the pebble falls

$$x = \frac{1}{2} gt^2 = \frac{1}{2} \left(9.8 \text{ m / s}^2\right)\left(16 \text{ s}\right)^2 = 1.25 \text{ km } \square$$

To think about refining this approximate answer, look up sound in the index to see that a typical value for the speed of sound in air is 343 m/s.

The time t_s it takes the sound to return a distance $x = 1.25$ km can be obtained using $x = v_s t_s$, where $a = 0$.

Therefore,

$t_s = x/v_s = (1250 \text{ m})/(343 \text{ m/s}) = 3.66$ s.

Thus, part of the 16 s is occupied by the sound returning, the stone falls for less than 16 s, and the well is less than 1.25 km deep.   Our next approximation could come from solving for x in

$$t_{fall} + t_s = 16 \text{ s} = \sqrt{\frac{2 x}{g}} + \frac{x}{v_s}$$

CHAPTER 2

PROBLEMS

2.3   **PROBLEM:**   The displacement versus time for a certain particle moving along the x axis is shown in Figure 2.14.   Find the average velocity in the time intervals (a) 0 to 2 s, (b) 0 to 4 s, (c) 2 s to 4 s, (d) 4 s to 7 s, (e) 0 to 8 s.

SOLUTION:

(a) $x = 0$ at $t = 0$ and $x = 10$ m at $t = 2$ s.

$$\overline{v} = \frac{Dx}{Dt} = \frac{10 \text{ m} - 0}{2 \text{ s} - 0} = 5.00 \text{ m/s} \ \square$$

(b) $x = 5$ m at $t = 4$ s.

$$\overline{v} = \frac{\Delta x}{\Delta t} = \frac{5 \text{ m} - 0}{4 \text{ s} - 0} = 1.25 \text{ m/s} \ \square$$

(c) $\overline{v} = \dfrac{\Delta x}{\Delta t} = \dfrac{5 \text{ m} - 10 \text{ m}}{4 \text{ s} - 2 \text{ m}} = -2.50 \text{ m/s} \ \square$

(d) $\overline{v} = \dfrac{\Delta x}{\Delta t} = \dfrac{-5 \text{ m} - 5 \text{ m}}{7 \text{ s} - 4 \text{ s}} = -3.33 \text{ m/s} \ \square$

(e) $\overline{v} = \dfrac{\Delta x}{\Delta t} = \dfrac{0 - 0}{8 \text{ s} - 0} = 0.00 \text{ m/s} \ \square$

2.9   **PROBLEM:**   Find the instantaneous velocity of the particle described in Figure 2.14 at the following times: (a) $t = 1$ s, (b) $t = 3$ s, (c) $t = 4.5$ s, and (d) $t = 7.5$ s.

SOLUTION:

(a) The tangent at $t = 1$ s coincides with the straight segment from (0 s, 0 m) to (2 s, 10 m)

$$\text{So, } v = \left. \frac{\Delta x}{\Delta t} \right)_{\text{on tangent}} = \frac{10 \text{ m} - 0}{2 \text{ s} - 0} = 5.00 \text{ m/s } \square$$

(b) The tangent at $t = 3$ s lies on the segment of the graph from (2 s, 10 m) to (4 s, 5 m).

$$v = \frac{dx}{dt} = \left. \frac{\Delta x}{\Delta t} \right)_{\text{tangent}} = \frac{5 \text{ m} - 10 \text{ m}}{4 \text{ s} - 2 \text{ s}} = -2.50 \text{ m/s } \square$$

(c) The tangent at $t = 4.5$ s is horizontal so,

$$v = 0 = \frac{5 \text{ m} - 5 \text{ m}}{5 \text{ s} - 4 \text{ s}} = 0 \ \square$$

(d) At $t = 7.5$ s the tangent passes through (7 s, -5 m) and (8 s, 0 m)

$$v = \frac{0 \text{ m} - (-5 \text{ m})}{8 \text{ s} - 7 \text{ s}} = 5.00 \text{ m/s } \square$$

2.15   **PROBLEM:**   A particle moves along the x axis according to the equation $x = 2 + 3t - t^2$, where x is in meters and t is in seconds.   At $t = 3$ s, find (a) the position of the particle, (b) its velocity, and (c) its acceleration.

SOLUTION: With the position given by $x = 2 + 3t - t^2$, we can use the rules for differentiation to write down the velocity as a function of time: $v = 3 - 2t$, and the acceleration: $a = -2$. Now we can evaluate x, v, and a at $t = 3$ s:

   (a) $x = 2 + 3(3) - (3)^2 = 2.00$ m $\square$

   (b) $v = 3 - 2(3) = -3.00$ m/s $\square$

   (c) $a = -2.00$ m/s^2 $\square$

2.17   **PROBLEM:**   Figure 2.19 shows a graph of v versus t for the motion
       of a motorcyclist as he starts from rest and moves along the
       road in a straight line.   (a) Find the average acceleration for
       the time interval $t_0 = 0$ to $t_1 = 6$ s.   (b) Estimate the time at
       which the acceleration has its greatest positive value and the
       value of the acceleration at that instant.   (c) When is the
       acceleration zero?   (d) Estimate the maximum negative value of
       the acceleration and the time at which it occurs.

SOLUTION: (a) $\overline{a} = \dfrac{\Delta v}{\Delta t} = \dfrac{8 \text{ m / s} - 0}{6 \text{ s} - 0} = 1.33 \text{ m/s}^2$ □

   (b)   The graph slopes upward most steeply at about 3 s □.
         Its tangent here has slope approximately

         $\dfrac{6 \text{ m / s} - 2 \text{ m / s}}{4 \text{ s} - 2 \text{ s}} = 2 \text{ m/s}^2$ □

   (c)   The graph has horizontal tangent at $t = 6$ s □, and
         throughout the interval 10.4 s $< t < 12$ s □.

   (d)   The graph slopes down most steeply at about $t = 8$ s □.
         Its tangent here passes approximately through (7 s, 8 m)

         and (9 s, 5 m), so it has slope $\dfrac{5 \text{ m} - 8 \text{ m}}{9 \text{ s} - 7 \text{ s}} = -1.5 \text{ m/s}^2$ □

2.23 **PROBLEM:** A jet plane lands with a speed of 100 m/s and can accelerate at a maximum rate of -5.0 m/s^2 as it comes to rest. (a) From the instant the plane touches the runway, what is the minimum time needed before it can come to rest? (b) Can this plane land on a small tropical island airport where the runway is 0.80 km long?

SOLUTION:

(a) Photograph the original point when plane first touches runway and the final point when it stops moving. Then $a = -5$ m/s^2 is constant between them. We have $v_0 = 100$ m/s and $v = 0$, so $v = v_0 + at$ gives

$$t = (v - v_0)/a = (0 - 100 \text{ m/s})/(-5 \text{ m/s}^2) = 20.0 \text{ s } \square$$

(b) $x - x_0 = \dfrac{1}{2}(v_0 + v)t = \dfrac{1}{2}(100 \text{ m/s} + 0)(20 \text{ s}) = 1000 \text{ m}$ so

the plane cannot land $\square$ on an 800 m runway.

Note we could also compute the stopping distance without using the time from part (a), using

$$v^2 = v_0^2 + 2a(x - x_0).$$

$$x - x_0 = \frac{v^2 - v_0^2}{2a} = \frac{0 - (100 \text{ m/s})^2}{2(-5 \text{ m/s}^2)} = +1000 \text{ m}$$

2.33 **PROBLEM:** Until recently, the world's land speed record was held
by Colonel John P. Stapp, USAF. On March 19, 1954, he road a
rocket-propelled sled that moved down the track at 632 mi/h. He
and the sled were safely brought to rest in 1.4 s. Determine
(a) the negative acceleration he experienced and (b) the
distance he traveled during this negative acceleration.

SOLUTION:

(a) We model the acceleration as constant during the
1.4 s stopping process, between

$$v_0 = 632 \text{ mi/h} = 632 \frac{\text{mi}}{\text{h}} \left(\frac{1609 \text{ m}}{1 \text{ mi}} \right) \left(\frac{1 \text{ h}}{3600 \text{ s}} \right) = 282 \text{ m/s}$$

and $v = 0$. Then $v = v_0 + at$ gives

$$a = \frac{v - v_0}{t} = \frac{0 - 282 \text{ m/s}}{1.4 \text{ s}} = -202 \text{ m/s}^2 \ \square$$

(b) $x - x_0 = \dfrac{1}{2} \left(v_0 + v \right) t = \dfrac{1}{2} (282 \text{ m/s} + 0) 1.4 \text{ s} = 198 \text{ m} \ \square$

2.43 **PROBLEM:** A daring cowboy sitting on a tree limb wishes to drop vertically onto a horse galloping under the tree. The speed of the horse is 10 m/s, and the distance from the limb to the saddle is 3 m. (a) What must be the horizontal distance between the saddle and limb when the cowboy makes his move? (b) How long is the cowboy in the air?

SOLUTION: We do part (b) first. Consider the vertical motion of the cowboy from leaving the limb (with $x_0 = 0$ at $x_0 = 3$ m) until reaching the saddle (at $x = 0$). We find his time of fall from

$$x - x_0 = v_0 t + \frac{1}{2}at^2$$

$$0 - 3 \text{ m} = 0 + \frac{1}{2}(-9.8 \text{ m}/\text{s}^2)t \qquad t = 0.782 \text{ s} \ \square$$

(a) In this time the horse travels horizontally according to

$v_0 = v = 10$ m/s, $a = 0$,

$x - x_0 = v_0 t = (10 \text{ m/s})(0.782 \text{ s}) = 7.82 \text{ m} \ \square$, so the cowboy must let go when the horse is this far away.

2.45 **PROBLEM:** The position of a softball tossed vertically upward is described by the equation $y = 7t - 4.9t^2$, where y is in meters and t is in seconds. Find (a) the initial speed v_0 at $t_0 = 0$, (b) the velocity at $t = 1.26$ s, and (c) The acceleration of the ball.

SOLUTION: From $y = 7t - 4.9t^2$ we can find the vertical velocity at all times by differentiating:

$v_y = \dfrac{dy}{dt} = 7 - 9.8t$; then the acceleration is

$a_y = \dfrac{dv_y}{dt} = -9.8$ m/s^2

(a) At $t = 0$, $v_y = 7 - 9.8(0) = 7.00$ m/s $\square$

(b) At $t = 1.26$ s, $v_y = [7 - (9.8)(1.26)] = -5.35$ m/s $\square$

(c) At all times after release, $a_y = -9.80$ m/s^2 $\square$

2.47 **PROBLEM:** A "superball" is dropped from a height of 2 m above the ground. On the first bounce the ball reaches a height of 1.85 m, where it is caught. Find the velocity of the ball (a) just as it makes contact with the ground and (b) just as it leaves the ground on the bounce. (c) Neglecting the time the ball spends in contact with the ground, find the total time required for the ball to go from the dropping point to the point where it is caught.

SOLUTION: The motion of the ball consists of free fall with acceleration $-g$; then contact with the floor for a short time with a large upward acceleration; and then free fall with $a = -g$. Because the acceleration changes, we cannot apply the four familiar equations to the whole motion, but only to constant-acceleration segments of it.

(a) For the downward fall we take $x_0 = 2$ m, $x = 0$, $v_0 = 0$.

Then $v^2 = v_0^2 + 2a(x - x_0)$ gives

$v^2 = 0^2 + 2(-9.8 \text{ m/s}^2)(0 - 2 \text{ m}) = +39.2 \text{ m}^2/\text{s}^2$

$v = 6.26$ m/s is unphysical; the velocity of impact is downward, -6.26 m/s $\square$. The time of fall comes from

$x - x_0 = v_0 t + \frac{1}{2}at^2$

$-2 \text{ m} = 0 + \frac{1}{2}(-9.8 \text{ m/s}^2)t$

$t^2 = 0.408 \text{ s}^2$

Here -0.639 s is unphysical, and $t = 0.639$ s.

(b) For the upward flight, $x_0 = 0$, $x = 1.85$ m, $v = 0$.

Then $v^2 = v_0^2 + 2a(x - x_0)$

$v_0 = \sqrt{0^2 + 2(-9.8 \text{ m/s}^2)(1.85 \text{ m})} = 6.02 \text{ m/s}$ $\square$

$x - x_0 = \frac{1}{2}(v_0 + v)t$

$1.85 \text{ m} - 0 = \frac{1}{2}(6.02 \text{ m/s} + 0)t \qquad t = 0.614 \text{ s}$

From (a) and (b), the total time in motion is

$0.639 \text{ s} + 0.614 \text{ s} = 1.25 \text{ s}$ $\square$

2.57   **PROBLEM:**   In a 100-m race, Maggie and Judy cross the finish line
in a dead heat, both taking 10.2 s.   Accelerating uniformly,
Maggie took 2.0 s and Judy 3.0 s to attain maximum speed, which
they maintained for the rest of the race.   (a) What was the
acceleration of each sprinter?   (b) What were their repective
maximum speeds?   (c) Which sprinter was ahead at the 6-s mark
and by how much?

SOLUTION: Maggie moves with constant positive acceleration a_M
for 2 s, and then with constant zero acceleration for 8.2 s,

covering altogether distance $x_{M1} + x_{M2}$ where $x_{M1} = \frac{1}{2}a_M(2 \text{ s})^2$ and

$x_{M2} = v_M(8.2 \text{ s})$, where v_M is her maximum speed.   We also have

$v_M = 0 + a_M(2 \text{ s})$, so by substitution

$$100 \text{ m} = \frac{1}{2}a_M(2 \text{ s})^2 + a_M(2 \text{ s})(8.2 \text{ s}) = (18.4 \text{ s}^2)a_M$$

and $a_M = 5.43 \text{ m/s}^2$ $\square$

Similarly for Judy, 100 m $= x_{J1} + x_J$ with

$$x_{J1} = \frac{1}{2}a_J(3 \text{ s})^2 \qquad v_J = a_J(3 \text{ s}) \qquad x_{J2} = v_J(7.2 \text{ s})$$

$$100 \text{ m} = \frac{1}{2}a_J(3 \text{ s})^2 + a_J(3 \text{ s})(7.2 \text{ s}) = 26.1 \text{ s}^2 \, a_J$$

$$a_J = 3.83 \text{ m/s}^2 \quad \square$$

(a) Their speeds after accelerating we have already written as

$v_M = a_M(2 \text{ s}) = 5.43 \text{ m/s}^2(2 \text{ s}) = 10.9 \text{ m/s}$ $\square$ and

$v_J = a_J(3 \text{ s}) = 3.83 \text{ m/s}^2(3 \text{ s}) = 11.5 \text{ m/s}\square$

(b) In the first 6 s, Maggie covers distance

$$\frac{1}{2} a_M(2 \text{ s}) + v_M(4 \text{ s}) = \frac{1}{2}5.43 \text{ m} / \text{s}^2(2 \text{ s})^2 + 10.9 \text{ m} / \text{s}(4 \text{ s}) = 54.3 \text{ m}$$

and Judy has moved

$$\frac{1}{2} a_J(3 \text{ s})^2 + v_J(3 \text{ s}) = \frac{1}{2} 3.83 \text{ m} / \text{s}^2(3 \text{ s})^2 + 11.5 \text{ m} / \text{s}(3 \text{ s}) = 51.7 \text{ m}$$

So Maggie $\square$ is ahead by 54.3 m - 51.7 m = 2.62 m $\square$

2.61 **PROBLEM:** Two objects, A and B, are connected by a rigid rod that has a length L. The objects slide along perpendicualr guide rails, as shown in Figure 2.23. If A slides to the left with a constant speed v, find the velocity of B when $\alpha = 60°$.

SOLUTION: The distances x and y are always related by $x^2 + y^2 = L^2$. Differentiating this equation with respect to time, we have $2x\dfrac{dx}{dt} + 2y\dfrac{dy}{dt} = 0$. Now $\dfrac{dy}{dt}$ is the unknown velocity of B while $\dfrac{dx}{dt} = -v$ is known: $\dfrac{dy}{dt} = +\dfrac{x}{y}v$

When $\alpha = 60°, \dfrac{y}{x} = \tan 60°$ and $\dfrac{dy}{dt} = \dfrac{v}{\tan 60°} = \dfrac{v}{\sqrt{3}}$

2.63   **PROBLEM:**   A particle undergoes a varying acceleration.   The
       velocity is measured at 0.5-s intervals and is tabulated as
       follows.   (a) Determine the average acceleration in each
       interval.   (b) Use a numerical integration procedure to
       determine the position of the particle at the end of each time
       interval.   Assume the initial position of the particle is zero.

t(s)	0	0.5	1.0	1.5	2.0	2.5	3.0	3.5	4.0	4.5	5.0
v(m/s)	0	1	3	4.5	7.0	9.5	10.5	12	14	15	17.5

SOLUTION: (a) For the interval from $t = 0$ to 0.5 s,

$$\overline{a} = \frac{\Delta v}{\Delta t} = \frac{1 \text{ m/s} - 0}{0.5 \text{ s} - 0} = 2 \text{ m/s}^2$$

For the interval from 0.5 s to 1 s,

$$\overline{a} = \frac{\Delta v}{\Delta t} = \frac{3 \text{ m/s} - 1 \text{ m/s}}{1 \text{ s} - 0.5 \text{ s}} = 4 \text{ m/s}^2$$

The other values are computed similarly:

t(s)	0	0.5	1	1.5	2	2.5	3	3.5	4	4.5	5
v(m/s)	0	1	3	4.5	7	9.5	10.5	12	14	15	17.5
Δv(m/s)		1	2	1.5	2.5	2.5	1	1.5	2	1	2.5
a(m/s^2)		2	4	3	5	5	2	3	4	2	5

2.63 (cont.)

(b) For $0 < t < 0.5$ s, the average speed is

$$\frac{1}{2}(0 + 1 \text{ m/s}) = 0.5 \text{ m/s}$$ so the step in distance is

$\Delta x = v\Delta t = (0.5 \text{ m/s})(0.5 \text{ s}) = 0.25$ m where we assume the acceleration is nearly constant over so short a time interval. Similarly, for 0.5 s $< y < 1$ s,

$$\Delta x \cong \frac{1}{2}(v_0 + v)\Delta t = \frac{1}{2}(1 \text{ m/s} + 3 \text{ m/s})(0.5 \text{ s}) = 1 \text{ m},$$

so at $t = 1$ s, it has moved to $x = 0 + 0.25$ m $+ 1$ m $= 1.25$ m. We continue,

t(s)	0	0.5	1	1.5	2	2.5	3	3.5	4	4.5	5
v(m/s)	0	1	3	4.5	7	9.5	10.5	12	14	15	17.5
v(m/s)	0.5		2	3.75	5.75	8.25	10	11.25	13	14.5	16.25
Δx(m)	0.25		1	1.9	2.9	4.1	5	5.6	6.5	7.2	8.1
v(m)	0	0.25	1.25	3.12	6	10.1	15.1	20.8	27.2	34.5	42.6

CHAPTER 3

QUESTIONS

15   **QUESTION:** As a projectile moves in its parabolic path, is there
any point along its path where the velocity and acceleration are
(a) perpendicular to each other? (b) parallel to each other?

ANSWER:

(a) At the top of its flight, a projectile has **v** horizontal and **a**

vertical. This is the only point where velocity and

acceleration are perpendicular.

(b) If the object is thrown straight up or down, then **v** and **a**

will be parallel throughout the downward motion. Otherwise,

velocity and acceleration are never parallel, but **v** gets

closer and closer to vertically downward as time goes on

during the downward motion.

19   **QUESTION:** A ball is tossed upwaard in the air by a passenger on
a train that is moving with a constant velocity. Describe the
path of the ball as seen by the passenger. Describe its path as
seen by a stationary observer outside the train. How would these
observations change if the train were accelerating along the
track?

ANSWER: The passenger sees the ball moving vertically up and
then down. An external observer sees the ball moving on a
parabola, with constant horizontal velocity equal to that of the
train, as well as contantly changing vertical velocity. If the
train were accelerating, say, horizontally forward, the passenger
would see the ball accelerating horizontally backward as well as
vertically downward. Relative to him, the ball would move on a
parabola with its axis inclined to the vertical. The outside
observer would see pure projectile motion for the ball, with
$a_x = 0$ and $a_y = -g$, the ball moving throughout its flight with v_x
the speed the train had at the moment of the ball's release.

22   **QUESTION:** An ice skater is executing a "figure eight,"
consisting of two equal, tangent circular paths. Throughout the
first loop she increases her speed uniformly, and during the
second loop she moves at a constant speed. Make a sketch of her
acceleration vector at several points along the path of motion.

ANSWER:

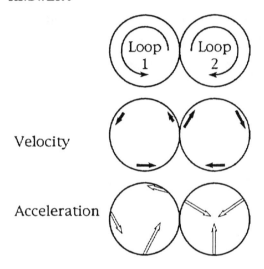

Figure 3.22

27   **QUESTION:**   A coin on a table is given an initial horizontal
     velocity such that it ultimately leaves the end of the table and
     hits the floor.   At the instant the coin leaves the end of the
     table, a ball is released from the same height and falls to the
     floor.   Explain why the two objects hit the floor simultaneously,
     even though the coin has an initial velocity.

     ANSWER: The two objects hit the floor simultaneously because
     their x and y motions are <u>independent</u>.   The fact that the coin
     has an initial x-velocity does not change its y-velocity.   Both
     objects fall with the same acceleration, g.

CHAPTER 3

PROBLEMS

3.3 **PROBLEM:** A motorist drives south at 20 m/s for 3 min, then turns west and travels at 25 m/s for 2 min, and finally travels north-west at 30 m/s, for 1 min. For this 6-min trip, find (a) the net vector displacement of the motorist, (b) the mortorist's average speed, and (c) the average velocity of the motorist.

SOLUTION: Her displacements total

$$\Delta \mathbf{r} = \left(20\,\frac{m}{s}\right)(3\ min)\left(\frac{60\ s}{1\ min}\right)south + \left(25\,\frac{m}{s}\right)(120\ s)\,west$$

$$+ \left(30\frac{m}{s}\right)(60\ s)\ northwest$$

Choosing $\mathbf{i}$ = east and $\mathbf{j}$ = north, we have

$$\Delta \mathbf{r} = (3.60\ km)(-\mathbf{j}) + (3.0\ km)(-\mathbf{i}) + (1.8\ km\ \cos 45°)(-\mathbf{i})$$

$$+ (1.8\ km\ \sin 45°)(\mathbf{j})$$

$$= (3 + 1.27)km(-\mathbf{i}) + (1.27 - 3.6)km\ \mathbf{j}$$

$$= (-4.27\mathbf{i} - 2.33\mathbf{j})\ km \quad \square$$

The answer can also be written as

$$\Delta \mathbf{r} = \sqrt{(-4.27\ km)^2 + (-2.33\ km)^2}\ at\ Arctan\ \frac{2.33}{4.27}\ south\ of\ west$$

$$= 4.87\ km\ at\ 209°\ from\ east. \quad \square$$

(b) The total distance or path-length traveled is
        $(3.60 + 3 + 1.8)\ km = 8.4\ km$

$$So\ average\ speed\ = \frac{8.41\ km}{6\ min}\left(\frac{1\ min}{60\ s}\right)\left(\frac{1000\ m}{km}\right) = 23.3\ m/s\ \square$$

(c) $\mathbf{v}_{av} = \dfrac{\Delta \mathbf{r}}{t} = \dfrac{4.87\ km\ at\ 209°}{360\ s} = 13.5\ m/s\ at\ 209°$

or, $\mathbf{v}_{av} = \dfrac{\Delta \mathbf{r}}{t} = \dfrac{(-4.27\ east - 2.33\ north)\ km}{360\ s}$

$\mathbf{v}_{av} = (11.9\ west + 6.47\ south)\ m/s$

3.5 **PROBLEM:** At $t = 0$, a particle moving in the xy plane with
     constant acceleration has a velocity of $\mathbf{v}_0 = (3\mathbf{i} - 2\mathbf{j})$ m/s at
     the origin.   At $t = 3$ s, the particle's velocity is
     $\mathbf{v} = (9\mathbf{i} + 7\mathbf{j})$ m/s.   Find (a) the acceleration of the particle and
     (b) its coordinates at any time t.

SOLUTION: $\mathbf{a} = \dfrac{\mathbf{v} - \mathbf{v}_0}{\Delta_t} = \dfrac{(9\mathbf{i} + 7\mathbf{j})\,\text{m} / \text{s} - (3\mathbf{i} - 2\mathbf{j})\,\text{m} / \text{s}}{3\ \text{s} - 0}$

$\mathbf{a} = (2.00\mathbf{i} + 3.00\mathbf{j})\ \text{m} / \text{s}^2$ □

(b) We choose the origin to be the particle's location at time
     zero.   Then since $\mathbf{a}$ is constant, we have

$\mathbf{r} = 0 + \mathbf{v}_0 t + \frac{1}{2}\mathbf{a}t^2$

$= 3\mathbf{i} - 2\mathbf{j})\,\text{m} / \text{s}\ t + \frac{1}{2}(2\mathbf{i} + 3\mathbf{j})\,\text{m} / \text{s}^2\ t$

$= (3t\ \text{m/s} + t^2\ \text{m/s}^2)\mathbf{i} + (-2t\ \text{m/s} + 1.5t^2\ \text{m/s}^2)\mathbf{j}$

3.11 **PROBLEM:** In a local bar, a customer slides an empty beer mug on the counter for a refill. The bartender is momentarily distracted and does not see the mug, which slides off the counter and strikes the floor 1.4 m from the base of the counter. If the height of the counter is 0.86 m, (a) with what velocity did the mug leave the counter, and (b) what was the direction of the mug's velocity just before it hit the floor?

SOLUTION: Choose the original point when the mug just leaves the counter. Here its motion is horizontal, with $v_{y0} = 0$. We choose this location to be the origin. Choose the final point just before it hits the floor. Here $x = 1.4$ m, $y = -0.86$ m, and v_x and v_y are both unknown. A list of knowns and unknowns looks like this:

For horizontal motion: $x = 1.4$ m, $v_x = v_{xo} = ?$, $a_x = 0$

For vertical motion: $y = -0.86$ m, $v_{y0} = 0$, $v_y = ?$, $a_y = -9.80$ m/s^2

(a) To find v_x in $x = v_x t$, we need first to know the time of flight. Since it is the one quantity in common between horizontal and vertical motions, we find it from

$$y = v_{y0}t + \frac{1}{2}a_y t \qquad -0.86 \text{ m} = 0 + \frac{1}{2}(-9.80 \text{ m}/\text{s}^2)t$$

At $t = 0.419$ s, $v_x = \dfrac{x}{t} = \dfrac{1.4 \text{ m}}{0.419 \text{ s}} = 3.34$ m/s

So, $v_0 = (3.34\mathbf{i} + 0\mathbf{j})$ m/s □

(b) $v_y = v_{y0} + a_y t = 0 - (9.8 \text{ m/s}^2)(0.419 \text{ s}) = -4.11$ m/s

Its impact angle is Arctan $\dfrac{-4.11 \text{ m}/\text{s}}{3.34 \text{ m}/\text{s}} = -50.9°$ □

What did you do wrong? Computing the elevation angle between initial and final points, Arctan (1.4/0.86) is a waste of time, since the straight line has nothing to do with the trajectory. The mug moves horizontally at the start, but NOT vertically at the end: you cannot guess the answer.

3.17   **PROBLEM:**   A place kicker must kick a football from a point 36 m (about 40 yards) from the goal, and the ball must clear the crossbar, which is 3.05 m high.   When kicked, the ball leaves the ground with a speed of 20.0 m/s at an angle of 53° to the horizontal.   (a) By how much does the ball clear or fall short of clearing the crossbar?   (b) Does the ball approach the crossbar while still rising or while falling?

SOLUTION: You cannot use $R = v_0^2 \dfrac{\sin(2\theta_0)}{g}$ to solve for any-thing, since ground and crossbar are at different elevations. Think instead.   Choose the origin and $t = 0$ where the ball just leaves the foot. Here $v_{x0} = v_x = (20 \text{ m/s})(\cos 53°) = 12$ m/s and $v_{y0} = (20 \text{ m/s}) \sin 53°$.   Choose the final point where the ball is vertically above or below the crossbar.   Here $x = 36$ m.   We want to know y (whether it is more or less than 3.05 m) and v_y (whether it is positive or negative.)

Choose $x = v_{0x}t$ to find $t = x/v_{x0} = 36$ m/(12 m/s) $= 2.99$ s.

Then $y = v_{y0}t + \dfrac{1}{2}a_y t^2$

$$y = (20 \text{ m/s}) \sin 53°(2.99 \text{ s}) + \frac{1}{2}(-9.8 \text{ m/s}^2)(2.99 \text{ s})$$

$= 47.8$ m $- 43.8$ m $= 3.94$ m,   so the ball clears □ by

$\Delta y = (3.94 - 3.05)$ m $= 0.89$ m □

(b) $v_y = v_{y0} + a_y t = (20 \text{ m/s}) \sin 53° - (9.8 \text{ m/s}^2)(2.99 \text{ s})$

$\quad v_y = -13.3$ m/s, so the ball is falling □.

3.27   **PROBLEM:**   An athlete roates a 1-kg discus along a circular path of radius 1.06 m.   The maximum speed of the discus is 20 m/s. Determine the magnitude of the maximum radial acceleration of the discus.

SOLUTION: The maximum radial acceleration occurs when maximum tangential speed is attained, just before the discus is released.

Here $a_r = v^2/r = (20 \text{ m/s})^2/(1.06 \text{ m}) = 377 \text{ m/s}^2$ □

3.31   **PROBLEM:**   Figure 3.19 represents the total acceleration of a particle moving clockwise in a circle of radius 2.5 m at a given instant of time.   At this instant, find (a) the centripetal accelertion, (b) the speed of the particle, and (c) its tangential acceleration.

SOLUTION: The acceleration has an inward component $a\cos30° = (15 \text{ m/s}^2)\cos30° = 13.0 \text{ m/s}^2$ □ -- this is the answer to part (a) -- and a forward component $a\sin30° = (15 \text{ m/s}^2)\sin30° = 7.50 \text{ m/s}^2$ □ -- this is the answer to (c).   The object is both rounding the curve and speeding up. (b) To find its speed right now, we use

$$a_c = v^2/r \qquad v = \sqrt{a_c r} = \sqrt{(13.0 \text{ m}/\text{s}^2)(2.5 \text{ m}} = 5.70 \text{ m/s} \text{ □}$$

3.39   **PROBLEM:**   The pilot of an airplane notes that the compass indicates a heading due west.   The speed of the airplane relative to the air is 150 km/h.   If there is a wind of 30 km/h toward the north, find the velocity of the airplane relative to the ground.

SOLUTION: The nose of the plane points west, so the velocity of plane relative to air is 150 km/h west.   The velocity of air relative to ground is 30 km/h north, so the velocity of the plane relative to the ground is

$\mathbf{V} = \mathbf{v}_{pg} = \mathbf{v}_{pa} + \mathbf{v}_{ag}$

   $= 150$ km/h west $+ 30$ km/h north

   $= \sqrt{150^2 + 30^2}$ km / at Arctan $\dfrac{30}{150}$ N of W.

$\mathbf{V} = 153$ km/h at $11.3°$ N of W $\square$

3.43   **PROBLEM:**   At $t = 0$ a particle leaves the origin with a velocity of 6 m/s in the positive y direction.   Its acceleration is given by $\mathbf{a} = (2\mathbf{i} - 3\mathbf{j})$ m/s^2.   When the particle reaches its *maximum* y coordinate, its y component of velocity is zero.   At this instant, find (a) the velocity of the particle and (b) its x and y coordinates.

SOLUTION: The object is not a projectile, but we can take its motion apart into components:

x-motion: $v_{xo} = 0$ y-motion: $v_{y0} = 6$ m/s

            $a_x = 2$ m/s^2 $a_y = -3$ m/s^2

            $v_x = ?$ $v_y = 0$

            $x = ?$ $y = ?$

Where we choose original and final points at the origin and the maximum y coordinate respectively.

Now $v_y = v_{y0} + a_y t$ gives $t = \dfrac{v_y - v_{y0}}{a_y} = \dfrac{0 - 6 \text{ m / s}}{-3 \text{ m / s}^2} = 2$ s

so $v_x = v_{x0} + a_x t = 0 + (2 \text{ m/s}^2)(2 \text{ s}) = 4 \text{ m/s}$
and $\mathbf{v} = (4 \text{ m/s})\mathbf{i} + 0\mathbf{j}$ $\square$

(b) $x = v_{xo}t + \dfrac{1}{2} a_x t^2 = 0 + \dfrac{1}{2}(2 \text{ m / s}^2)(2 \text{ s})^2 = 4$ m

   $y = v_{y0}t + \dfrac{1}{2} a_y t = (6 \text{ m / s})(2 \text{ s}) + \dfrac{1}{2}\left(-3 \text{ m / s}^2\right)(2 \text{ s})^2 = 6$ m

so, $\mathbf{r} = (4.00\mathbf{i} + 6.00\mathbf{j})$m $\square$

3.45   **PROBLEM:** A batter hits a pitched baseball 1 m above the ground, imparting to it a speed of 40 m/s. The resulting line drive is caught on the fly by the left fielder 60 m from home plate, with his glove 1 m above the ground. If the shortstop, 45 m from home plate and in line with the drive, were to jump straight up to make the catch instead of allowing the left fielder to make the play, how high above the ground would his glove have to be?

SOLUTION: We FIRST take o and f points at the batter and the left fielder, to find the original angle with the horizontal. Note y stands for difference in height between o and f points, so

$$y = 0; \quad v_{y0} = (40 \text{ m/s}) \sin\theta_0; \quad a_y = -9.8 \text{ m/s}^2$$
$$x = 60 \text{ m}; \quad v_{x0} = (40 \text{ m/s}) \cos\theta_0; \quad a_x = 0$$

Now, $y = v_{y0}t + \frac{1}{2}a_y t^2$ gives

$$0 = (40 \text{ m/s}) \sin\theta_0 t + \frac{1}{2}(-9.8 \text{ m/s}^2)t$$

The root $t = 0$ represents when ball left bat; the time of flight is the other root, $t = 2(40 \text{ m/s}) \sin\theta_0 / (9.8 \text{ m/s}^2)$ substituting this into

$$x = v_{x0}t = (40 \text{ m/s}) \cos\theta_0 (2)(40 \text{ m/s}) \sin\theta_0 / (9.8 \text{ m/s}^2) = 60 \text{ m}$$

gives $2(\cos\theta_0)(\sin\theta_0) = \sin 2\theta_0 = \dfrac{60 \text{ m})(9.8 \text{ m/s}^2)}{(40 \text{ m/s})^2} = 0.368$

$2\theta_0 = 21.6°; \quad \theta_0 = 10.8°$

Now SECOND we take o and f points at the batter and when the ball passes above the shortstop:

$$y = ?; \quad v_{y0} = (40 \text{ m/s}) \sin 10.8°; \quad a_y = -9.8 \text{ m/s}^2$$
$$x = 45 \text{ m}; \quad v_{x0} = (40 \text{ m/s}) \cos 10.8°; \quad a_x = 0$$

3.45 (cont.)

Now, $x = v_{x0}t$ gives

$t = x/v_{x0}$ = 45 m/(40 m/s)(cos10.8°) = 1.15 s

So, $y = v_{y0}t + \frac{1}{2}a_y t^2$

$\qquad$ = (40 m/s)sin10.8°(1.15 s) − (4.9 m/s^2)(1.15 s)^2

$\qquad$ = 2.14 m

This is the extra elevation above starting point. Altogether the ball is 1 + 2.14 m = 3.14 m □ above the ground: the shortstop would have to reach as high as Michael Jordan.

3.51   **PROBLEM:** A boat requires 2 min to cross a river that is 150 m wide. The boat's speed relative to the water is 3 m/s, and the river current flows at a speed of 2 m/s. At what upstream or downstream points could the boat reach the opposite shore in 2 min?

SOLUTION: Choose the y-axis in the direction of water flow. The velocity of water relative to earth is $\mathbf{v}_{we} = (2 \text{ m/s})\mathbf{j}$, while we can write the velocity of boat relative to water as $\mathbf{v}_{bw} = (3 \text{ m/s})\cos\theta \; \mathbf{i} + (3 \text{ m/s})\sin\theta \; \mathbf{j}$, where θ is the bearing-angle of the boat away from pointing straight across. The x-component of motion is described by

$$(150 \text{ m})\mathbf{i} = (3 \text{ m/s})\cos\theta \; \mathbf{i}(120 \text{ s}); \; \theta = \text{Arccos} \frac{1.25}{3} = \pm 65.4°$$

Now if the boat heads downstream, $\theta = +65.4°$, and its total velocity is

$$\mathbf{v}_{be} = \mathbf{v}_{bw} + \mathbf{v}_{we}$$
$$= (3 \text{ m/s})\cos 65.4° \; \mathbf{i} + (3 \text{ m/s})\sin 65.4° \; \mathbf{j} + (2 \text{ m/s})\mathbf{j}$$
$$= (1.25 \text{ m/s})\mathbf{i} + (4.73 \text{ m/s})\mathbf{j}$$

So its displacement in 2 minutes is

$$\Delta\mathbf{r} = [(1.25 \text{ m/s})\mathbf{i} + (4.73 \text{ m/s})\mathbf{j}](120 \text{ s})$$
$$= (150 \text{ m})\mathbf{i} + (567 \text{ m})\mathbf{j}$$

If the boat heads upstream, $\theta = -65.4°$, similarly,

$$\mathbf{v}_{be} = (3 \text{ m/s})\cos(-65.4°)\mathbf{i} + (3 \text{ m/s})\sin(-65.4°)\mathbf{j} + (2 \text{ m/s})\mathbf{j}$$
$$= (1.25 \text{ m/s})\mathbf{i} + (0.727 \text{ m/s})\mathbf{j}$$

$$\Delta\mathbf{r} = \mathbf{v}_{be}\Delta t = (150 \text{ m})\mathbf{i} - (87.3 \text{ m})\mathbf{j}$$

So the boat could land 567 m downstream or 87.3 m upstream. □

3.65   **PROBLEM:** A hawk is flying horizontally at 10.0 m/s in a
straight line, 200 m above the ground. A mouse it has been
carrying is released from its grasp. The hawk continues on its
path at the same speed for 2 seconds before attempting to
retrieve its prey. To accomplish the retrieval, it dives in a
straight line at constant speed recaptures the mouse 3.0 m
above the ground. (a) Assuming no air resistance, find the
diving speed of the hawk. (b) What angle did the hawk make
with the horizontal during its descent? (c) For how long did
the mouse "enjoy" free fall?

SOLUTION: Our standard equations describe only constant-
acceleration motions, so we must consider three: the projectile
($\mathbf{a}$ = $0\mathbf{i}$ - $g\mathbf{j}$) motion of mouse, the $\mathbf{a}$ = 0 flight of the hawk
during her 2 s reaction time, and the $\mathbf{a}$ = 0 dive of the hawk.
Name your ignorance: call $\mathbf{v_h}$ = $v_{hx}\mathbf{i}$ + $v_{hy}\mathbf{j}$ the hawk's diving
velocity and t the time for which raptor and prey are
separated. Then for the mouse we have

y = -197 m; v_x = 10 m/s; v_{y0} = 0; a_y = -9.8 m/s^2; a_x = 0.
Thus, we can answer part (c):

$$y = v_{y0}t + \frac{1}{2}a_yt^2; \quad -197 \text{ m} = 0 + \frac{1}{2}(-9.8 \text{ m}/\text{s}^2)t^2$$

t = 6.34 s $\square$

Then x = v_xt = (10 m/s)(6.34 s) = 63.4 m

The hawk moves horizontally at 10 m/s for 2 s and at v_{hx} for
(6.34 - 2)s = 4.34 s, to reach the same point as the mouse:

$$63.4 \text{ m} = (10 \text{ m/s})(2 \text{ s}) + v_{hx}(4.34 \text{ s}); \quad v_{hx} = \frac{43.4 \text{ m}}{4.34 \text{ s}} = 10 \text{ m}/\text{s}$$

The hawk moves vertically not at all for 2 s and then at v_{hy}
for 4.34 s: -197 m = 0 + v_{hy}{4.34 s}; v_{hy} = -45.4 m/s

(a) Now $|\mathbf{v_h}|$ = $\sqrt{10^2 + (-45.4)^2}$ m/s = 46.5 m/s $\square$

(b) θ = Arctan $\dfrac{v_{hy}}{v_{hx}}$ = Arctan $\dfrac{-45.4 \text{ m}/\text{s}}{10 \text{ m}}$ = -77.6° $\square$

3.71 **PROBLEM:** A ball bearing is dropped from the point $x = 4$ m, $y = 2$ m. At the same moment, a second bearing is launched from $x = 0$, $y = 0$ at an angle of $20°$ above the positive x axis at a speed of 6 m/s. Determine (a) the minimum distance between the bearings and (b) the time at which this minimum occurs. *Suggestion:* If you cannot solve this problem analytically, you may wish to write and run a short computer program that locates the minimum of the *square* of the distance between the bearings.

SOLUTION: The position of the first ball relative to the origin is given by $\mathbf{r}_1 = (4 \text{ m})\mathbf{i} + (2 \text{ m})\mathbf{j} - \dfrac{1}{2}(9.8 \text{ m}/\text{s}^2)\mathbf{j}\ t$.

That of the second is

$\mathbf{r}_2 = 0\mathbf{i} + 0\mathbf{j} + (6 \text{ m/s} \cos 20°)t\ \mathbf{i} + (6 \text{ m/s} \sin 20°)t\ \mathbf{j}$

$\qquad - \dfrac{1}{2}(9.8 \text{ m}/\text{s}^2)\mathbf{j}\ t$

The distance between them is

$\Delta r = |\mathbf{r}_1 - \mathbf{r}_2|$

$\qquad = |(4 \text{ m} - 6 \text{ m/s} \cos 20°t)\mathbf{i} + (2 \text{ m} - 6 \text{ m/s} \sin 20°t)\mathbf{j}|$

Its square is

$(4 \text{ m} - 6 \text{ m/s} \cos 20°t)^2 + (2 \text{ m} - 6 \text{ m/s} \sin 20°t)^2$

$\Delta r) = 16 \text{ m}^2 - (48 \text{ m/s} \cos 20°)t + (36 \text{ m}^2/\text{s}^2 \cos^2 20°)t^2 + 4 \text{ m}^2$

$\qquad - (24 \text{ m/s})\sin 20°t + (36 \text{ m}^2/\text{s}^2)\sin^2 20°\ t^2$

$\qquad = 20 \text{ m}^2 - (53.3 \text{ m/s})t + (36 \text{ m}^2/\text{s}^2)t^2$

3.71 (cont.)

We can find the value of t to minimize this <u>either</u> by using a standard calculus technique, setting its derivative equal to zero:

$$\frac{d(\Delta r)^2}{dt} = -53.3 \text{ m/s} + (72 \text{ m}^2/\text{s}^2)t = 0 \qquad t = 0.740 \text{ s}$$

<u>or else</u> by directed trial and error, by substituting in values of t and homing in on the value that makes the expression the smallest:

At $t = 0$, $(\Delta r)^2 = 20 \text{ m}^2$

At $t = 2$ s, $(\Delta r)^2 = 57.4 \text{ m}^2$

At $t = 1$ s, $(\Delta r)^2 = 2.7 \text{ m}^2$

At $t = 0.5$ s, $(\Delta r)^2 = 2.35 \text{ m}^2$

At $t = 0.6$ s, and so on

Either method finds, to three digits, t = 0.740 s ☐ as the answer to (b). The minimum distance itself is

$$\sqrt{20 \text{ m}^2 - (53.3 \text{ m/s})(0.740 \text{ s}) + (36 \text{ m}^2/\text{s}^2)(0.74 \text{ s})^2} = 0.511 \text{ m} \ ☐$$

CHAPTER 4

QUESTIONS

6   **QUESTION:** The observer in the elevator of Example 4.8 would claim that the "weight" of the fist is T, the scale reading. This is obviously wrong. Why does this observation differ from that of a person outside the levator, at rest with respect to the elevator?

ANSWER: The observer in the elevator does not see the fish accelerating in his frame of reference. He imagines the fish to be in equilibrium, but Newton's laws do not apply in his non-inertial reference frame.

15   **QUESTION:** The force of gravity is twice as great on a 20-N rock as on a 10-N rock. Why doesn't the 20-N rock have a greater free-fall acceleration?

ANSWER: The 20 N rock has larger mass. With twice as much "inertia" in Latin or "pigheadedness" in English, it would move with only half the acceleration of a 10 N rock under the same force, and it requires just twice as much total force to move with the same acceleration.

16   **QUESTION:** Is it possible to have motion in the absence of a force? Explain.

ANSWER: Motion requires no force. Newton's first law says that motion needs no cause but continues by itself. The simplest motion to think of is that of a meteoroid in outer space (initated by a glider on an air track). An object feeling no forces moves and moves steadily.

17   **QUESTION:**   Is there any relation between the net force acting on an object and the direction in which it moves?   Explain.

ANSWER: There is no relation between the total force on an object and the direction of its current motion.   Forces describe what the rest of the universe does to the object, and the environment can push it forward, backward, sideways, or not at all.   On the other hand, if the total force is always in one direction, the velocity vector will generally turn closer and closer to that direction, over time.

19   **QUESTION:**   A 0.15-kg baseball is thrown upward with an initial speed of 20 m/s.   If air resistance is neglected, what is the net force on the ball (a) when it reaches half its maximum height? (b) when it reaches its peak?

ANSWER: At all points in its trajectory, not just at the halfway mark and apex, the total force, and the only force on the baseball is its weight, $\mathbf{w} = m\mathbf{g} = 1.47$ N down.

25   **QUESTION:**   If a small sports car collides head-on with a massive truck, which vehicle experiences the greater impact force?   Which vehicle experiences the greater acceleration?   Explain.

ANSWER: A strain gage or heavy-duty bathroom scale placed between the vehicles reads the same whichever way it faces.   Car and truck experience equal forces, each away from the other.   The smaller mass of the car makes it stop with much greater acceleration.

CHAPTER 4

PROBLEMS

4.5 **PROBLEM:** A 5.0-g bullet leaves the muzzle of a rifle with a speed of 320 m/s. What average force is exerted on the bullet while it is traveling down the 0.82-m-long barrel of the rifle? Assume the bullet's acceleration is constant.

SOLUTION: This is two problems: first we must find the bullet's acceleration and then the forces on it. For the trip down the barrel we have, $v_0 = 0$, $v = 320$ m/s,

$x - x_0 = 0.82$ m.

Then $v^2 = v_0^2 + 2a(x - x_0)$

$$a = \frac{v^2 - v_0^2}{2(x - x_0)} = \frac{(320 \text{ m} / \text{s})^2 - 0}{2(0.82 \text{ m})} = 62.4 \text{ km} / \text{s}$$

The bullet's weight is

$w = mg = (5 \times 10^{-3} \text{ kg})(9.8 \text{ m/s}^2) = 49.0$ mN

We suppose the barrel is horizontal. Then the barrel-floor exerts a normal force according to

$\Sigma F_y = ma_y$; $+n - 49$ mN $= 0$; $n = 49$ mN.

The forward force P of the exploding gunpowder is much larger:

$\Sigma F_x = ma_x$; $P = 5 \times 10^{-3}$ kg $(62.4 \times 10^3 \text{ m/s}^2)$; $P = 312$ N.

So the forces are 49.0 mN down by the Earth, 49.0 mN up by the barrel, and 312 N forward by the hot gases. □

4.9   **PROBLEM:**   A woman weighs 120 lb. Determine (a) her weight in
      newtons and (b) her mass in kilograms.

      SOLUTION: Her weight is 120 lb down.   She could change it
      either by eating submarine sandwiches or by moving to a differ-
      ent gravitational environment, such as French Guiana. Her weight

      is $\mathbf{w}$ = 120 lb down = (120 lb down)$\left(\dfrac{1 \text{ N}}{0.255 \text{ lb}}\right)$ = 533 N down $\square$

      (b) Her mass is the same throughout the Universe, so long as she
           refuses those subs.   We find it by thinking of her weight
           just here, where g = 9.80 m/s^2:

           $\mathbf{w}$ = m$\mathbf{g}$ down; $|\mathbf{w}|$ = mg

           $m = \dfrac{|\mathbf{w}|}{g} = \dfrac{533 \text{ N}}{9.8 \text{ m} / \text{s}^2} = 54.4$ kg $\square$

      [The variation of an object's weight over the Earth's surface
      was discovered in French Guiana, when fine pendulum clocks were
      carried there from Paris.   I own stock in the franchiser of your
      local sub shop.]

4.15 **PROBLEM:**   A boat moves through the water with two forces acting
      on it.   One is a 2000-N forward push by the motor; the other is
      an 1800-N resistive force due to the water.   (a) What is the
      acceleration of the 1000-kg boat?   (b) If it starts from rest,
      how far will it move in 10 s?   (c) What will be its speed at the
      end of this time?

      SOLUTION: Choose the x-axis forward.   Then the resistive force
      is (−1800 N)$\mathbf{i}$.

      We have $\Sigma F_x = ma_x$;   (2000 N)$\mathbf{i}$ − (1800 N)$\mathbf{i}$ = (1000 kg)$\mathbf{a}$

      $\mathbf{a}$ = (0.200 m/s^2)$\mathbf{i}$ $\square$

      (b) $x - x_0 = v_0 t + \dfrac{1}{2}at^2 = 0 + \dfrac{1}{2}(0.2 \text{ m} / \text{s}^2)(10 \text{ s}) = 10.0$ m $\square$

      (c) $\mathbf{v} = \mathbf{v}_0 + \mathbf{a}t = 0 + (0.2$ m/s$^2)\mathbf{i}$ (10 s) = (2.00 m/s)$\mathbf{i}$ $\square$

4.21 PROBLEM: A barefoot field-goal kicker imparts a speed of 30 m/s to a football initially at rest. If the football has a mass of 0.5 kg and the time of contact with the football is 0.025 s, what is the force exerted on the foot?

SOLUTION: $F = m\dfrac{\Delta v}{\Delta t}$

For a football of mass 0.5 kg, the force acting on the football (and also <u>felt</u> by the foot) is

$$F = 0.5 \text{ kg})\frac{(30 \text{ m/s})}{(0.025 \text{ s}} = 600 \text{ N} \ \square$$

4.27 PROBLEM: A 1-kg mass is observed to accelerate at 10 m/s^2 in a direction 30° north of east (Fig. 4.21). One of the two forces acting on the mass has a magnitude of 5 N and is directed north. Determine the magnitude and direction of the second force acting on the mass.

SOLUTION: Choose east = **i** and north = **j**. Notice how we substitute into both sides of $\Sigma \mathbf{F} = m\mathbf{a}$:

$\mathbf{F}_1 + \mathbf{F}_2 = (1 \text{ kg})(10 \text{ m/s}^2)$ at 30°

$\mathbf{F}_1 + (5 \text{ N})\mathbf{j} = (10 \text{ N})\cos 30° \ \mathbf{i} + (10 \text{ N})\sin 30° \ \mathbf{j}$

$\mathbf{F}_1 = (8.66 \text{ N})\mathbf{i} + (5 \text{ N})\mathbf{j} - (5 \text{ N})\mathbf{j} = (8.66 \text{ N})\mathbf{i} + (0 \text{ N})\mathbf{j}$

$\mathbf{F}_1 = 8.66 \text{ N east} \ \square$

4.28 **PROBLEM:** Find the tension in each cord of the systems decribed in Figure 4.22. (Neglect the mass of the cords.)

SOLUTION: (a)

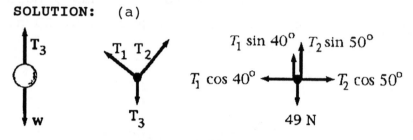

Figure 4.28(a)

First draw a free-body diagram of the object: We suppose it to be near the Earth, so $w = mg = $ (5 kg)(9.8 m/s^2) = 49 N. We suppose it to be in equilibrium, so $\Sigma F_y = ma_y$;

$+T_3 - 49$ N = 0; $T_3 = 49$ N □

Now choose the knot in the cords as object:

We draw the free-body diagram twice for clarity.

Now we have the simultaneous equations

$\Sigma F_x = ma_x$; $-T_1 \cos 40° + T_2 \cos 50° = 0$

$\Sigma F_y = ma_y$; $+T_1 \sin 40° - 49$ N $+ T_2 \sin 50° = 0$

We solve by substitution:

$T_2 = T_1 \cos 40°/\cos 50°$

$T_1 \sin 40° + (T_1 \cos 40°/\cos 50°)\sin 50° = 49$ N

$T_1 = $ (49 N)/(0.643 + 0.913) = 31.5 N □

$T_2 = $ (31.5 N)$\cos 40°/\cos 50°$ = 37.5 N □

4.28 (b) For the hanging mass,

$$w = mg = (10 \text{ kg})(9.8 \text{ m/s}^2) = 98 \text{ N}$$

$$\Sigma F_y = ma_y; \qquad T_3 - 98 \text{ N} = 0; \qquad T_3 = 98 \text{ N} \quad \square$$

For the knot, the free-body diagrams are

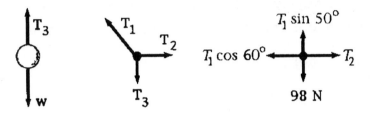

Figure 4.28(b)

$$\Sigma F_y = ma_y; \qquad T_1 \sin 60° - 98 \text{ N} = 0; \qquad T_1 = 113 \text{ N} \quad \square$$

$$\Sigma F_x = ma_x; \quad -T_1 \cos 60° + T_2 = 0; \; T_2 = (113 \text{ N})\cos 60° = 56.6 \text{ N} \quad \square$$

4.31 **PROBLEM:** A 0.15-kg baseball moving at 20 m/s strikes the glove of a catcher. The glove recoils a distance of 8 cm. What average force does the glove exert on the ball? What average force does the ball exert on the glove?

SOLUTION: To find the average acceleration, we model the stopping process as having constant acceleration. Let the 20 m/s have direction **i**. Then, $v^2 = v_0^2 + 2a(x - x_0)$ gives

$$a = \frac{v^2 - v_0^2}{2(x - x_0)} = \frac{0 - (20 \text{ m}/\text{s})^2}{2(0.08 \text{ m})} = -2500 \text{ m}/\text{s}$$

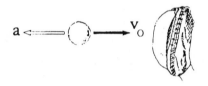

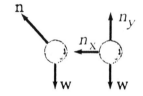

Figure 4.31

The glove exerts a normal force on the ball to support its weight and stop it:

$\Sigma F_y = ma_y;$ $+n_y - w = 0;$

$n_y = w = mg = (0.15 \text{ kg})(9.8 \text{ m/s}^2) = 1.47 \text{ N}$

$\Sigma F_x = ma_x;$ $-n_x = (0.15 \text{ kg})(-2500 \text{ m/s}^2) = -375 \text{ N}$

So the glove exerts $\mathbf{n} = (-375 \text{ N})\mathbf{i} + (1.47 \text{ N})\mathbf{j}$ ☐

The ball must then exert equal and opposite force on the glove:

$(+375 \text{ N})\mathbf{i} - (1.47 \text{ N})\mathbf{j}$ ☐

4.41 **PROBLEM:** A block is given an initial velocity of 5 m/s up a frictionless 20° incline (Fig 4.27). How far up the incline does the block slide before coming to rest?

SOLUTION: Every successful physics student (this means you) learns to solve inclined-plane problems. **Trick one:** take the x-axis along the incline. Then $a_y = 0$. **Trick two:** recognize that the 20° angle between x-axis and horizontal in the motion picture imples a 20° angle between weight and y-axis in the forces picture. Why? Because "angles are equal if their sides are perpendicular, right side to right side and left side to left side." Either you learned this theorem in geometry or you learn it now, since it is the theorem used most often in

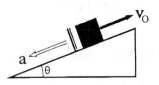

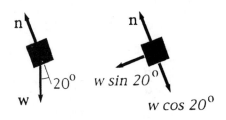

Figure 4.41

physics. **Trick three:** The 20° angle lies between **w** and y-axis, so the y-component of weight is w cos 20° and its x-component is $-w$ sin20°. Do not get the two backwards. Now, $\Sigma F_x = ma_x$; $-w\sin 20° = ma_x$. Substituting $w = mg$ gives $-mg \sin 20° = ma_x$. $a_x = (-9.8 \text{ m/s}^2) \sin20° = -3.35 \text{ m/s}^2$

[If we needed the normal force, it is given by $\Sigma F_y = ma_y$;

$+n - w \cos20° = 0$; $n = mg \cos20°$]

We analyze the motion according to $v_0 = 5$ m/s; $v = 0$;

$a = -3.35 \text{ m/s}^2$; $x - x_0 = ?$; $v^2 = v_0^2 + 2a(x - x_0$

$$x - x_0 = \frac{(v^2 - v_0^2)}{2a} = \frac{0 - (5 \text{ m / s})^2}{2(-3.35 \text{ m / s}^2)} = 3.73 \text{ m} \ \square$$

If the motion picture shows v_0 up the incline, why is there no force up the incline? Because we choose to think only of the motion after the block leaves your hand.

4.43 **PROBLEM:** A 3.0-kg mass is moving in a plane, with its x and y coordinates given by $x = 5t^2 - 1$ and $y = 3t^3 + 2$ (x and y are in meters and t is in seconds). Find the magnitude of the net force acting on this mass at $t = 2.0$ s.

SOLUTION: In Newton's theory of motion, it is not position, nor velocity, but only acceleration that is associated with the outside influences on the object.

From $x = 5t^2 - 1$; $y = 3t^3 + 2$; we differentiate to find the velocity components

$$v_x = \frac{dx}{dt} = 10t; \qquad v_y = \frac{dy}{dt} = 9t^2$$

and then again to find the acceleration

$$a_x = \frac{dv_x}{dt} = 10 \qquad a_y = \frac{dv_y}{dt} = 18t$$

The total force on the object must then be

$$\Sigma \mathbf{F} = m\mathbf{a} = (3 \text{ kg})(10\mathbf{i} + 18t\mathbf{j}) \text{ m/s}^2$$
$$= (30\mathbf{i} + 54t\mathbf{j}) \text{ N}$$

At $t = 2$, this is $\Sigma \mathbf{F} = (30\mathbf{i} + 108\mathbf{j})$ N, with

$$|\Sigma \mathbf{F}| = \sqrt{30^2 + 108^2} \text{ N} = 112 \text{ N} \quad \square$$

4.45 **PROBLEM:** A car of mass 1500 kg is being pulled up a loading ramp inclined at 30° with the horizontal, as in Figure 4.28. The car is attached to a cable, which passes over a frictionless pulley to a 10,000 N counterweight. Find (a) the tension in the cable, and (b) the acceleration of the system. (c) What mass should the counterweight have in order for the car to move down the incline at an acceleration of 2 m/s²? (Ignore any effects of friction.)

SOLUTION: The car's weight is
$w = mg = $ (1500 kg)(9.8 m/s²) $ = $ 14700 N.

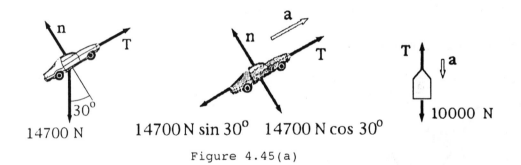

Figure 4.45(a)

Its component down the incline is (14700 N)sin30°, so for the car $\Sigma F_x = ma_x$ reads (−14700 N)sin30° + $T = $ (1500 kg)a. The counterweight moves the same number of centimeters in every second as the car. It moves down but with acceleration of the same magnitude: $\Sigma F_y = ma_y$;

$$T - 10,000\ N = \frac{10000\ N}{9.8\ m/s^2}(-a)$$

[Your most likely mistakes are to forget the minus sign in (−a) and to forget to find its mass

$$m = \frac{w}{g} = \frac{10000\ N}{9.8\ m/s^2} = 1020\ kg]$$

4.45 (cont.)

Now the two simultaneous equations are

$(-14700 \text{ N}) \sin 30° + T = (1500 \text{ kg}) a$

$T - 10000 \text{ N} = (1020 \text{ kg})(-a)$

We can solve by subtracting:

$(-14700 \text{ N}) \sin 30° + T - T + 10{,}000 \text{ N} = (1500 \text{ kg}) a + (1020 \text{ kg}) a$

$a = \dfrac{(10{,}000 \text{ N} - 7350 \text{ N})}{2520 \text{ kg}} = 1.05 \text{ m/s}^2 \ \square$

That is answer (b). The tension we can find from either of the $\Sigma F = ma$ equations:

$-7350 \text{ N} + T = (1500 \text{ kg})(1.05 \text{ m/s}^2)$

$T = 8.93 \text{ kN} \ \square$

(c) Now, for the car,

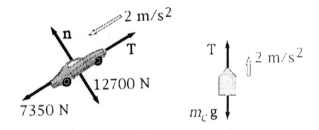

Figure 4.45(b)

$\Sigma F_x = ma_x; \quad -7350 \text{ N} + T = (1500 \text{ kg})(-2 \text{ m/s}^2); \quad T = 4350 \text{ N}.$
For the counterweight, be careful!

$\Sigma F_y = ma_y; \quad 4350 \text{ N} - m_c (9.8 \text{ m/s}^2) = m_c (+2 \text{ m/s}^2)$

$m_c = \dfrac{4350 \text{ N}}{11.8 \text{ m/s}^2} = 369 \text{ kg} \ \square$

4.54 **PROBLEM:**   Three blocks are in contact with each other on a frictionless, horizontal surface, as in Figure 4.32.   A horizontal force **F** is applied to m_1.   If m_1 = 2 kg, m = 3 kg, m_3 = 4 kg, and F = 18 N, find (a) the acceleration of the blocks, (b) the *resultant* force on each block, and (c) the magnitudes of the contact forces between the blocks.

SOLUTION: Let m_1 exert force P forward on m_2.   Then m_2 exerts force P backward on m_1.   Let m_2 exert force R forward on m_3. Then m_3 exerts R backward on m_2.   The free body diagrams look like this:

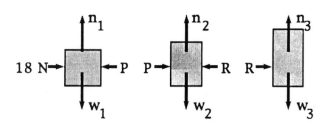

Figure 4.54

We write $\Sigma F_x = ma_x$ three times for the three objects:

$$+18 \text{ N} - P = (2 \text{ kg})a$$
$$+P - R = (3 \text{ kg})a$$
$$+R = (4 \text{ kg})a$$

Adding gives 18 N = (9 kg)a; a = 2.00 m/s^2 ☐

(b) Now $\Sigma F_{\text{on } 1}$ = 18 N - P = (2 kg)(2 m/s^2) = 4.00 N ☐

$\Sigma F_{\text{on } 2}$ = $P - R$ = (3 kg)(2 m/s^2) = 6.00 N ☐

$\Sigma F_{\text{on } 3}$ = R = (4 kg)(2 m/s^2) = 8.00 N ☐

(c) The contact force between m_2 and m_3 is R = 8.00 N ☐.

We find P from 18 N - P = 4 N. P = 14.0 N ☐

4.55 **PROBLEM:** A high diver of mass 70 kg jumps off a board 10 m above the water. If his downward motion is stopped 2 s after he enters the water, what average upward force did the water exert on the diver?

SOLUTION: We find his speed at first contact with the water by analyzing his free fall:

$$v^2 = v_0^2 + 2a(x - x_0) = 0 + 2(-9.8 \text{ m/s}^2)(-10 \text{ m})$$

$v = -14$ m/s   (we chose the negative root to describe his downward motion.)

Now that −14 m/s is the original velocity for his stopping motion:

$$v = v_0 + at$$

$$a = \frac{v - v_0}{t} = \frac{0 - (-14 \text{ m/s})}{2 \text{ s}} = +7 \text{ m/s}$$

While the diver stops, the water pushes up with force R:

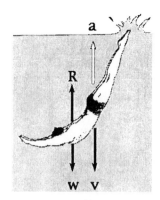

Summing forces, $\Sigma F_y = ma_y$ and

$$+R - (70 \text{ kg})(9.8 \text{ m/s}^2) = (70 \text{ kg})(7 \text{ m/s}^2)$$

$$R = (70 \text{ kg})(16.8 \text{ m/s}^2) = 1180 \text{ N} \; \square$$

4.57 **PROBLEM:** Two forces, $\mathbf{F}_1 = (-6\mathbf{i} - 4\mathbf{j})$ N and $\mathbf{F}_2 = (-3\mathbf{i} + 7\mathbf{j})$ N, act on a particle of mass 2 kg that is initially at rest at coordinates (-2 m, +4 m). (a) What are the components of the particle's velocity at $t = 10$ s? (b) In what direction is the particle moving at $t = 10$ s? (c) What displacement does the particle undergo during the first 10 s? (d) What are the coordinates of the particle at $t = 10$ s?

SOLUTION: We suppose that $\mathbf{F}_1$ and $\mathbf{F}_2$ are the only forces. We ignore the object's weight. Then we find its acceleration vector:

$\Sigma \mathbf{F} = m\mathbf{a}$

$(-6\mathbf{i} - 4\mathbf{j})$ N $+ (-3\mathbf{i} + 7\mathbf{j})$ N $= (2$ kg$)\mathbf{a}$

$\mathbf{a} = \dfrac{(-9\mathbf{i} + 3\mathbf{j})\,\text{N}}{2\,\text{kg}} = (-4.5\mathbf{i} + 1.5\mathbf{j})\ \text{m/s}^2$

(a) Now its final velocity:

$\quad \mathbf{v} = \mathbf{v}_0 + \mathbf{a}t = 0 + (-4.5\mathbf{i} + 1.5\mathbf{j})\text{m/s}^2(10\ \text{s})$

$\quad\quad = (-45\mathbf{i} + 15\mathbf{j})\ \text{m/s}\ \square$

(b) Direction $= $ Arctan $\dfrac{15\ \text{m/s}}{-45\ \text{m/s}} = 162°\ \square$

(c) $\Delta\mathbf{r} = \mathbf{v}_0 t + \dfrac{1}{2}\mathbf{a}t$

$\quad\quad = \quad + \dfrac{1}{2}\left(-4.51\mathbf{i} + 1.5\mathbf{j}\right)\text{m / s}^2\left(10\ \text{s}\right)^2$

$\quad\quad = (-225\mathbf{i} + 75\mathbf{j})\text{m}\ \square$

(d) It starts from $\mathbf{r}_0 = (-2$ m$)\mathbf{i} + (4$ m$)\mathbf{j}$, so

$\quad\quad \mathbf{r}_f = \mathbf{r}_0 + \Delta\mathbf{r} = (-2$ m$)\mathbf{i} + 4$ m$)\mathbf{j} - (225$ m$)\mathbf{i} + (75$ m$)\mathbf{j}$

$\quad\quad = (-227\mathbf{i} + 79\mathbf{j})\text{m}\ \square$

CHAPTER 5

QUESTIONS

5   **QUESTION:**   The driver of a speeding empty truck slams on the
     brakes and skids to a stop through a distance *d*. (a) If the
     truck's mass were doubled by a heavy load, what would be the
     truck's "skidding distance"?   (b) If the initial speed of the
     truck were halved, what would be the truck's "skidding distance"?

ANSWER:

(a) Doubling the truck's mass would double the normal force the
    road exerts and double the backward force of kinetic
    friction. With twice the total force acting on twice the
    mass, its acceleration would be the same and so would all
    other parameters about its motion, including distance.

(b) With the same acceleration and a smaller original speed, the
    time required to stop would be cut in half, and moving with
    one-half the average speed for one-half the time, the truck
    would travel one-quarter the distance. We can also see this
    from the proportionality of x to v_0^2, rather than v_0, in the

    equation $x = \dfrac{(v^2 - v_0^2)}{2a}$

8   **QUESTION:**   Because the Earth rotates about its axis and about the Sun, it is a noninertial frame of reference.   Assuming the Earth is a uniform sphere, why would the *apparent weight* of an object be greater at the poles than at the equator?

ANSWER: Hang a fish on a scale at the north or south pole. The tension in the spring equals the weight of the fish, supporting it in equilibrium, so the scale reads the full force of the Earth's on the fish.   Now repeat the experiment at the equator.   The fish is not in equilibrium, but moves in a circle with the Earth's rotation and has centripetal acceleration

$$a_c = a_c = \frac{v^2}{r} = \frac{1}{r}\left(\frac{2\pi r}{T}\right)2 = \frac{4\pi^2 r}{T^2} = \frac{4\pi^2(6.37 \times 10^6 \text{ m})}{(86,400 \text{ s})^2}$$

$a_c = 0.0337$ m/s^2 down.

To provide a corresponding downward total force on the fish, $\Sigma F = ma_c$, its weight must exceed the spring tension.   Thus the spring tension, which is the apparent weight of the fish, is less than its actual weight by its mass times this 33.7 mm/s^2. (The revolution of the earth around the sun involves a smaller

centripetal acceleration, $\dfrac{4\pi^2(1.5 \times 10^{11} \text{ m})}{(3.16 \times 10^7 \text{ s})^2} = 5.93$ mm/^{2} toward the

sun, the same size at poles and equator.   The sun's gravity provides the force to cause this acceleration, so it does not affect the apparent weight of the fish.)

21   **QUESTION:**   Consider a sky diver falling through air *before*
      reaching terminal speed.   As the speed of the sky diver
      increases, what happens to her acceleration?

      ANSWER: The sky diver feels constant downward weight.   As her
      downward speed increases, she experiences an increasing upward
      air resistance force, always less than her weight.   The sum of
      weight downward and increasing resistance upward gives decreasing
      downward total force, so her acceleration decreases.

29   **QUESTION:**   An object executes circular motion with a constant
      speed whenever a net force of constant magnitude acts perpendi-
      cular to the velocity.   What happens to the speed if the force is
      not perpendicular to the velocity?

      ANSWER: An object can move in a circle even if the total force
      on it is not perpendicular to its velocity; but then its speed
      will change.   Resolve the total force into an inward radial
      component and a perpendicular tangential component. If the
      tangential force is forward, the object will speed up, and it
      will lose speed if the tangential component acts backward on it.

CHAPTER 5

PROBLEMS

5.5   **PROBLEM:** A block moves up a 45° incline with constant speed
under the action of a force of 15 N applied *parrallel* to the
incline. If the coefficient of kinetic friction is 0.3,
determine (a) the weight of the block and (b) the minimum force
required to allow the block to move *down* the incline at constant
speed.

SOLUTION:

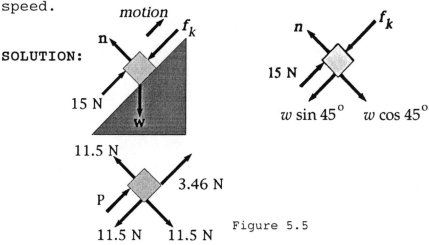

Figure 5.5

Review in chapter 4 about how to resolve the weight into
components parallel to the incline (x-axis) and perpendicular
to the motion (y-axis).

Now $\Sigma F_y = ma_y$ gives $+n - w\cos 45° = 0$; $n = w\cos 45°$.

Thus $f_k = \mu_k n = 0.3w\cos 45°$

and $\Sigma F_x = ma_x$ gives $+15$ N $- w\sin 45- 0.3\ w\cos 45° = 0$

$$w = \frac{15 \text{ N}}{\sin 45°} + 0.3\cos 45° = 16.3 \text{ N} \ \square$$

5.5   (b) As the object slides down the plane, it experiences the
           force $w \sin 45°$ down the incline, amounting to
           $w \sin 45° = (16.3N)\sin 45° = 11.5$ N. Friction, which opposes
           the motion, will be up along the incline,
           $\mu_k n = (0.3)(16.3\cos 45°) = 3.46$ N. Then the extra applied
           force P must be up the incline to prevent acceleration:

           $\Sigma F_x = ma_x$

           $+P - 11.5N + 3.46\ N = 0$

           $P = 8.08$ N up the incline $\square$

5.13 **PROBLEM:** A woman at an airport is towing her 20-kg suitcase at
      constant speed by pulling on a strap at an angle of θ above the
      horizontal (Fig. 5.13). She pulls on the strap with a 35-N
      force, and the friction force on the suitcase 20 N. (a) What
      angle does the strap make with the horizontal? (b) What normal
      force does the ground exert on the suitcase?

      SOLUTION: The forces on the suitcase are its weight
      $w = mg = (20$ kg$)(9.8$ m/s$^2) = 196$ N down,
      the unknown normal force n up, the pull
      $(35$ N$)\cos\theta$ forward and $(35$ N$)\sin\theta$ up, and
      friction 20 N backward. Moving at
      constant speed, the case is in
      equilibrium and the acceleration is zero:

      $\Sigma F_x = ma_x$ gives -20 N $- (35$ N$)\cos\theta = 0$

      $\theta = $ Arccos$(0.571) = 55.2°$ $\square$

      $\Sigma F_y = ma_y$ reads

      $+n + (35$ N$) \sin 55.2° - 196$ N $= 0$

      $n = 167$ N $\square$

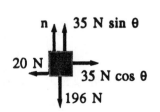

                                                        Figure 5.13

5.21 **PROBLEM:** A crate of eggs is positioned in the middle of the
flatbed of a truck as the truck negotiates a curve in the road.
The curve may be regarded as an arc of a circle of radius 35 m.
If the coefficient of static friction between the crate and the
flatbed of the truck is 0.6, what must be the maximum speed of
the truck if the crate is not to slide during the maneuver?

SOLUTION: Call the mass of the egg
crate m. The forces on it are its
weight mg vertically down, the
normal force n of the truck bed
vertically up, and static friction
directed to oppose relative sliding
motion of the crate over the truck
bed. The friction force is directed
radially inward. It is the only hori-
zontal force on the crate, so it must provide the centripetal
acceleration. When the truck has maximum speed, friction

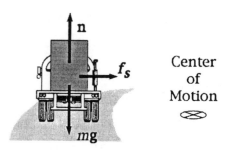

Figure 5.21

$f_s \leq \mu_s n$ will have its maximum value $f_s = \mu_{is} n$.

$\Sigma F_y = ma_y$ gives $+n - mg = 0$ $\qquad n = mg$

So $\Sigma F_x = ma_x$ reads $f_s = ma_c$

$\mu_s n = mv^2/r$

$\mu_s mg = mv^2/r$

The mass divides out.

$v = \sqrt{\mu_s gr}$

$v = \sqrt{0.6 \times 9.8 \text{ m} / \text{s}^2 \times 35 \text{ m}}$

$v = 14.3 \text{ m/s } \square$

5.29   **PROBLEM:**   A 40-kg child sits in a conventional swing of length
       3 m, supported by two chains.   If the tension in each chain is
       350 N when the swing is at its lowest point, find (a) the
       child's speed at the lowest point and (b) the force of the seat
       on the child at the lowest point.
       (Neglect the mass of the seat.)

       SOLUTION:

       (a)   Choose seat-and-child-together as one
             object. The forces on it are weight
             $w = mg = (40 \text{ kg})(9.8 \text{ m/s}^2) = 392 \text{ N}$
             down, and two 350 N forces up.

             Now $\Sigma F_y = ma_y$ becomes

             $+350 \text{ N} + 350 \text{ N} - 392 \text{ N}$

             $= (40 \text{ kg})(v^2/3 \text{ m})$

             $v = \sqrt{(308 \text{ N})(3 \text{ m} / 40 \text{ kg}}$

             $v = 4.81 \text{ m/s}$ $\square$

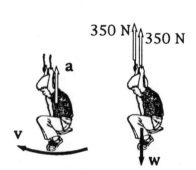

Figure 5.29

       (b)   The child alone feels an upward normal force from the seat,
             and 392 N of downward weight:

             $\Sigma F_y = ma_y$

             $+n - 392 \text{ N} = (40 \text{ kg})(4.81 \text{ m/s})^2/(3 \text{ m})$

             $n = 700 \text{ N}$ $\square$

5.37 **PROBLEM:**   The driver of a motor boat cuts the engine when the
speed is 10 m/s, and the boat coasts to rest. The equation
governing the motion of the motorboat during this period is
$v = v_0 e^{-ct}$, where v is the speed at time t, v_0 is the inital
speed, and c is a constant. At $t = 20$ s the speed is 5 m/s.
(a) Find the constant c. (b) What is the speed at $t = 40$ s?
(c) Differentiate the preceding expression for $v(t)$ and thus
show that the acceleration of the boat is proportional to the
speed at any time.

SOLUTION:

(a)   We must fit the equation $v = v_0 e^{-ct}$ to the two data points:
at $t = 0$, $v = 10$ m/s:

Substitution gives 10 m/s = $v_0 e^0$ = v_0 x 1 so v_0 = 10 m/s.

And at $t = 20$ s, $v = 5$ m/s: substitution gives

5 m/s = 10 m/s $e^{-c \text{ x } 20 \text{ s}}$

$0.5 = e^{-c \text{ x } 20 \text{ s}}$

ln 0.5 = $(-c)(20$ s$)$

$$c = \frac{-\ln 0.5}{20 \text{ s}} = 0.0347/\text{s} \quad \square$$

(b) At all times $v = (10$ m/s$)e^{-0.0347 \, t/s}$

At $t = 40$ s, $v = (10$ m/s$)e^{-0.347 \text{ x } 40} = 2.50$ m/s $\square$

(c) The acceleration is the rate-of-change of velocity:

$$a = \frac{dv}{dt} = \frac{d}{dt}v_0 e^{-ct} = v_0\left(e^{-ct}\right)(-c)$$

$= -c(v_0 e^{-ct}) = -cv = -0.0347 \ v/\text{s}$

Thus the acceleration is a negative constant times the
speed.

5.41 **PROBLEM:** A hailstone of mass 4.8×10^{-4} kg falls through the air and experiences a net force given by $F = -mg + Dv^2$ where $D = 2.5 \times 10^{-5}$ kg/m. (a) Calculate the terminal speed of the hailstone. (b) Use Euler's method of numerical analysis to find the speed and position of the hailstone at 0.2 s intervals, taking the initial speed to zero. Continue the calculation until the hailstone reached 99% of terminal speed.

SOLUTION:

(a) At terminal speed the acceleration and total force are zero: $-mg + Dv^2 = 0$

$$v = \sqrt{\frac{mg}{D}} = \sqrt{(4.8 \times 10^{-4} \text{ kg}) \frac{(9.80 \text{ m} / \text{s}^2)}{(2.5 \times 10^{-5} \text{ kg} / \text{m})}}$$

$v = -13.7$ m/s

We choose the negative root for downward velocity

(b) At $t = 0$ we take $x = 0$. It starts from rest, so $v = 0$ and

$\Sigma F = -mg + Dv^2 = -(4.8 \times 10^{-4} \text{ kg})(9.80 \text{ m/s}^2) + 0$

so $a = \Sigma F/m = -9.80$ m/s^2. Now at $t = 0.2$ s,

$x_{new} \cong x_{old} + v_{old}\Delta t = 0 + 0 = 0$

$v_{new} \cong v_{old} + a_{old}\Delta t = 0 - (9.8 \text{ m/s}^2)(0.2 \text{ s}) = -1.96$ m/s

Now iterate: at $t = 0.2$ s

$\Sigma F = -mg + Dv^2$

$\quad = (-4.70 \times 10^{-3} \text{ N}) + (2.5 \times 10^{-5} \text{ kg/m})(-1.96 \text{ m/s})^2$

$\quad = -4.61$ mN

So $a = \dfrac{\Sigma F}{m} = -9.60$ m/s^2.

At $t = 0.4$ s, $x \cong 0 + (-1.96 \text{ m/s})(0.2 \text{ s}) = -0.392$ m

$v \cong -1.96 \text{ m/s} - (9.60 \text{ m/s}^2)(0.2 \text{ s}) = -3.88$ m/s

So far we have filled in the first two-and-one-half lines of this table, which we continue:

5.41 con't

t, s	x, m	v, m/s	ΣF, mN	a, m/s^2
0	0	0	−4.70	−9.80
0.2	0	−1.96	−4.61	−9.60
0.4	−0.392	−3.88	−4.33	−9.02
0.6	−1.17	−5.68	−3.90	−8.12
0.8	−2.30	−7.31	−3.37	−7.02
1.0	−3.77	−8.71	−2.81	−5.85
1.2	−5.51	−9.88	−2.26	−4.72
1.4	−7.48	−10.8	−1.78	−3.70
1.6	−9.65	−11.6	−1.36	−2.84
1.8	−12.0	−12.1	−1.03	−2.14
2.0	−14.4	−12.6	−0.762	−1.59

...now listing results after each fifth step

3.0	−27.4	−13.5	−0.154	−0.321
4.0	−41.0	−13.7	−0.0291	−0.0606
5.0	−54.7	−13.7	−0.00542	−0.0113

The hailstone never attains terminal speed exactly, but passes
99% of it at 3.5 seconds, and 99.99% at 6.0 seconds.

5.43   **PROBLEM:**   A 50-kg parachutist jumps from an airplane and falls
to Earth with a drag force proportional to the square of the
speed, $R = Dv^2$.   Take $D = 0.2$ kg/m (with the parachute closed)
and $D = 20$ kg/m (with the chute open).   (a) Determine the
terminal speed of the parachutist in both configurations, before
and after the chute is opened.   (b) Set up a numerical analysis
of the motion and compute the speed and position as functions of
time, assuming the jumper begins the descent at 1000 m above the
ground and is in free fall for 10 s before opening the
parachute.   (*Hint:* When the parachute opens, a sudden large
acceleration takes place; a smaller time step may be necessary
in this region.)

SOLUTION:

(a) With constant velocity, $\Sigma F = 0$, $-mg + Dv_t^2 = 0$

$$v_t = \sqrt{\frac{mg}{D}} = -\sqrt{\frac{(50 \text{ kg}) (9.8 \text{ m}/\text{s}^2)}{0.2 \text{ kg}/\text{m}}} = -49.5 \text{ m}/\text{s with chute}$$

closed and

$$v_t = -\sqrt{\frac{(50 \text{ kg})(9.8 \text{ m}/\text{s}^2)}{20 \text{ kg}/\text{m}}} = -4.95 \text{ m}/\text{s with chute open.}$$

5.43 (b)   We can try a 0.2 s time interval for $0 < t < 10$ s, then
0.02 s for 10 s $< t < 11$ s, and then back to 0.2 s.   We
calculate first $a = (-50 \text{ kg} \times 9.8 \text{ m/s}^2 + Dv^2)/50 \text{ kg}$,

time, s	height, m	velocity, m/s	acceleration, m/s^2
0	1000	0	-9.80
0.2	1000	-1.95	-9.80
0.4	999.6	-3.92	-9.78
1	996	-9.71	-9.56
4	931	-33	-5.75
7	814	-44	-2.21
10	676	-47.8	-0.70 or +900
10.1	673	-16.9	107
11	667	-5.10	0.618
50	474	-4.95	0

then $x_{new} = x_{old} + v_{old}\Delta t$ and $v_{new} = v_{old} + a\Delta t$.
Some of the lines of output are

The parachutist reaches the ground after 145 seconds.

5.47 **PROBLEM:** 1 3-kg mass hangs at one end of a rope that is attached to a support on a railroad car. When the car accelerates to the right, the cord makes an angle of 4° with the vertical, as shown in Figure 5.47 (a). Find the acceleration of the car.

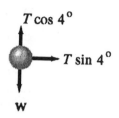

Figure 5.47

SOLUTION: We choose to use the inertial frame of the ground. With its string making a constant angle, the mass has the same acceleration as the car, horizontal in direction. The forces on it are its weight,

$w = mg = $ (3 kg)(9.80 m/s^2) = 29.4 N, and the rope tension, with components $T \sin 4°$ forward and $T \cos 4°$ up. (There is no normal force since it does not touch the floor or wall.)

Then $\Sigma F_y = ma_y$ gives $+ T \cos 4° - 29.4$ N $= 0$

$T = 29.5$ N and $\Sigma F_x = ma_x$ reads

$+ 29.5 \sin 4° = $ (3 kg) a

$a = 0.685$ m/s^2 □

The value of the mass is unnecessary information, since it divides out in the symbolic solution

$a = T \sin 4°/m = (mg/\cos 4°)(\sin 4°/m) = g \tan 4°.$

5.51 **PROBLEM:** A spinning ball of radius 5.0 cm slows uniformly in 0.3 s from 30 rev/min to rest. Compute the radial tangential, and net accelerations of a point on the equator of the ball at the beginning of this time period.

SOLUTION: A point on the equator moves in a circle of circumference $2\pi r = 2\pi(5 \text{ cm}) = 31.4$ cm, so its original speed is

$$v = 30\frac{\text{rev}}{\text{min}}\left(\frac{31.4 \text{ cm}}{1 \text{ rev}}\right)\left(\frac{1 \text{ min}}{60 \text{ s}}\right) = 15.7 \text{ cm/s.}$$

Its radial acceleration is

$a_c = v^2/r = (15.7 \text{ cm/s})^2/5 \text{ cm}$

$a_c = 49.3 \text{ cm/s}^2$ inward $\square$

Its tangential acceleration is backward

$a_t = \Delta v/\Delta t = (0 - 15.7 \text{ cm/s})/0.3 \text{ s}$

$a_t = -52.4 \text{ cm/s}^2$ $\square$

These are the perpendicular components of its total acceleration, of magnitude $a_n = \sqrt{(49.3)^2 + (52.4)^2}$ cm / s $= 71.9 \text{ cm/s}^2$, and direction inward and backward at angle

Arctan $\frac{52.4}{49.3} = 46.7^c$ to the radius:

$a_{net} = 71.9 \text{ cm/s}^2$ inward and $46.7°$ backward $\square$

5.57 **PROBLEM:** An engineer wishes to design a curved exip ramp for a toll road in such a way that a car will not have to rely on friction to round the curve without skidding. Suppose that a typical car rounds the curve with a speed of 30 mi/h (13.4 m/s) and the radius of the curve is 50 m. At what angle should the curve be banked? (See Fig.5.57)

SOLUTION: Figure 5.57 shows all the forces on the car. Do not treat this like an inclined-plane problem! You know the magnitude of the car's acceleration,

$a_c = v^2/r = (13.4 \text{ m/s})^2/50 \text{ m} = 3.59 \text{ m/s}^2$;   Figure 5.57

its direction is horizontally inward toward the center. So take the x-axis horizontal and the y-axis vertical. Leave the weight alone and resolve the normal force into $n \cos \theta$ upward and $n \sin \theta$ inward horizontally.

Now $\Sigma F_y = ma_y$ gives $+n \cos \theta - mg = 0$

$$n = mg/\cos \theta$$

and $\Sigma F_x = ma_x$ reads $-n \sin \theta = -m(3.59 \text{ m/s}^2)$.

Substitution gives $(mg/\cos \theta)\sin \theta = m \ 3.59 \text{ m/s}^2$

$\tan \theta = (3.59 \text{ m/s}^2)/(9.8 \text{ m/s}^2) = 0.366$

$\theta = 20.1° \ \square$

5.59 **PROBLEM:** A model airplane of mass
0.75 kg flies in a horizontal circle at
the end of a 60-m control wire, with a
speed of 35 m/s. Compute the tension in
the wire if it makes a constant angle of
20° with the horizontal. The airplane is
acted upon by the tension in the control

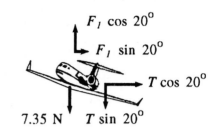

Figure 5.59

line, its weight, and the aerodynamic lift, which act at 20°
inward from the vertical, as shown in Figure 5.25.

SOLUTION: The plane's acceleration is toward the center of the
circle of motion, so it is horizontal. The radius of the circle
of motion is $(60 \text{ m}) \cos 20° = 56.4$ m, and the size of the

acceleration is $a_c = \dfrac{v^2}{r} = \dfrac{(35 \text{ m / s})^2}{56.4 \text{ m}} = 21.7 \text{ m/s}^2$.

Now with $w = mg = (0.75 \text{ kg}) (9.8 \text{ m/s}^2) = 7.35$ N, the forces
resolve as shown.

Therefore, $\Sigma F_x = ma_x$ and $\Sigma F_y = ma_y$ read, respectively,

$-T \cos 20° - F_1 \sin 20° = (-0.75 \text{ kg}) (21.7 \text{ m/s}^2) = -16.3$ N

$+F_1 \cos 20° - T \sin 20° - 7.35 \text{ N} = 0$

These are two simultaneous equations in two unknowns. We can
solve by substitution:

$F_1 = (16.3 \text{ N} - T \cos 20°)/\sin 20°$

$(16.3 \text{ N} - T \cos 20°) \cos 20°/\sin 20° - T \sin 20° = 7.35$ N

$(16.3 \text{ N}) \cos 20° - T \cos^2 20° - T \sin^2 20° = 7.35 \text{ N} \sin 20°$

Now $\cos^2 \theta + \sin^2 \theta = 1$, so

$T = (16.3 \text{ N}) \cos 20° - (7.35 \text{ N}) \sin 20° = 12.8 \text{ N}$ $\square$

5.61   PROBLEM: An amusement park ride consists of a large vertical cylinder that spins about its axis fast enough that any person inside is held up against the wall when the floor drops away (Fig. 5.61).   The coefficient of static friction between the person and the wall is μ_s, and the radius of the cylinder is R. (a) Show that the *maximum* period of revolution necessary to keep the person from falling is $T = (4\pi^2 R \mu_s/g)^{1/2}$.   (b) Obtain a numerical value for T if $R = 4$ m and $\mu_s = 0.4$.   How many revolutions per minute does the cylinder make?

SOLUTION:

(a)   The normal force of the wall pushes inward:

$$N = \frac{MV^2}{R} = \frac{m}{R}\left(\frac{2\pi R}{T}\right)^2 = \frac{4\pi^2 Rm}{T^2}$$

and the frictional force balances the person's weight:

$$f_s = \mu_s n = mg$$

Therefore, with $\mu_s n = mg$,

$$\mu_s n = \mu_s \frac{4\pi^2 Rm}{T^2} = mg, \text{ and}$$

$$T^2 = \frac{4\pi^2 R\mu_s}{g}$$

or $T = \sqrt{\dfrac{4\pi^2 R\mu_s}{g}}$, as required.

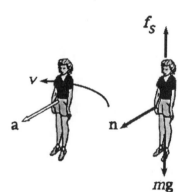

Figure 5.61

5.61 (b) $T = \left[\dfrac{\left(4\pi^2\right)\left(4 \text{ m}\right)\left(0.4\right)}{\left(9.8 \text{ m / s}^2\right)}\right]^{1/2} = 2.54 \text{ s } \square$

The angular speed is $\dfrac{1 \text{ rev}}{2.54 \text{ s}} = 23.6 \text{ rev/min } \square$

Think carefully about this problem. Why is the normal force inward? Because the wall is vertical and on the outside. Why is there no upward normal force? Because there is no floor. Why does friction act up? To oppose the possible relative motion of the person sliding down the wall. Why is it not kinetic friction? Because person and wall are moving together, stationary with respect to each other. And why is there no outward force on the person? Because no other object pushes outward on her. She pushes outward on the wall as the wall pushes inward on her with a force of π.

CHAPTER 6

QUESTIONS

1   **QUESTION:**   Estimate the gravitational force between you and a person 2 m away from you.

ANSWER: To make an order-of-magnitude estimate, we choose typical values for each piece of data we need.   In the answer we think only of the order of magnitude, the closest power of ten, as meaningful.   Suppose one person has mass 60 kg and the other 70 kg.   Each feels the same force of attraction toward the other:

$$F_g = \frac{GM_1M_2}{r^2} = \frac{(6.67 \times 10^{-11} \text{ N} \cdot \text{m}^2)(60 \text{ kg})(70 \text{ kg})}{(\text{kg}^2)(2 \text{ m})^2} = 7 \times 10^{-8} \text{ N}$$

or on the order of 10^{-7} N.□

3   **QUESTION:**   If someone told you that astronauts are weightless in orbit because they are beyond the pull of gravity, would you accept this statement?   Explain.

ANSWER: Astronauts in orbit around the Earth are held in orbit by the force of the Earth's gravity.   They float around freely inside their spacecraft because they and the craft are in free fall together.   No matter which side of the spacecraft we take as the floor, there is no normal force on the astronauts because the whole system is "falling" toward Earth.   The Earth's gravitational field gets weaker farther away, but there is no boundary beyond which it is zero.

7 **QUESTION:** Is it possible for an electric field to exist in empty space? Explain.

ANSWER: An electric field exists in empty space. The electric field is defined to be that influence that crosses from one charged object to another charged object across the empty space between them. In this way, the electric force acts on each charge.

12 **QUESTION:** At a given instant, a proton moves in the positive x direction in a region where there is a magnetic field in the negative z direction. What is the direction of the magnetic force? Will the proton continue to move in the positive x direction? Explain.

ANSWER: Call the $+x$ velocity direction to the right and the $-z$ magnetic field direction away from you. Then the magnetic force direction is perpendicular to both, and is <u>up</u> according to the right-hand rule with your hand held thus:

. Then the proton will feel this upward $q\mathbf{v} \times \mathbf{B}$ force

and its path will curve upward, in the y-direction.

CHAPTER 6

PROBLEMS

6.1 **PROBLEM:** Two identical, isolated particles, each of mass 2 kg, are separated by a distance of 30 cm. What is the magnitude of the gravitational force of one particle on the other?

SOLUTION: Each object attracts the other with force

$$F_g = \frac{GM_1M_2}{r^2} = \frac{6.67 \times 10^{-11} \text{ N} \cdot \text{m}^2(2 \text{ kg})(2 \text{ kg})}{\left(\text{kg}^2\right)(0.3 \text{ m})^2} = 2.96 \times 10^{-9} \text{ N} \;\square$$

6.7 **PROBLEM:** Plaskett's binary system consists of two stars that revolve in a circular orbit about a center of gravity midway between them. This means that the masses of the two stars are equal (Fig 6.15). If the orbital velocity of each star is 220 km/s and the orbital period of each is 14.4 days, find the mass M of each star. (For comparison, the mass of our Sun is 1.99×10^{30} kg.)

SOLUTION: The circumference of the stars' common orbit is

$$2\pi r = vT = 220 \times 10^3 \text{ m/s})(14.4 \text{ d})\left(\frac{86,400 \text{ s}}{1 \text{ d}}\right)$$

$$2\pi r = 2.74 \times 10^{11} \text{ m}$$

so each is distant from the center by

$$r = 2.74 \times 10^{11} \text{ m}/2\pi = 4.36 \times 10^{10} \text{ m},$$

while the distance between the stars is $2r = 8.72 \times 10^{10}$ m. The gravitational force of one star on the other supplies the centripetal force about the center:

$$\Sigma F = ma \text{ reads } \quad \frac{GMM}{(2r)^2} = \frac{Mv^2}{r}$$

$$M = \frac{v^2 4r}{G} = \frac{\left(220 \times 10^3 \text{ m/s}\right)^2(4)\left(4.36 \times 10^{10} \text{ m}\right)}{6.67 \times 10^{-11} \text{ N} \cdot \text{m}^2/\text{kg}^2} = 1.26 \times 10^{32} \text{ kg} \;\square$$

(or about 63 solar masses for each star).

6.9   **PROBLEM:** Suppose that 1 g of hydrogen is separated into 6 x 10^{23} electrons and 6 x 10^{23} protons. Suppose also that the protons are placed at the Earth's north pole and the electrons are placed at the south pole. What is the resulting compressional force on the Earth?

SOLUTION: The charge of the 6 x 10^{23} protons is (6 x 10^{23})(1.60 x 10^{-19} C) = 9.6 x 10^4 C. The electrons have negative charge of the same magnitude. The distance between them is to be the Earth's diameter, 2 x 6.37 x 10^6 m = 1.27 x 10^7 m. So each attracts the other with force of magnitude

$$F_e = k_e \frac{|q_1||q_2|}{r^2} = 8.99 \times 10^9 \ \frac{N \cdot m^2}{C^2} \frac{(9.6 \times 10^4 \ C)^2}{(1.27 \times 10^7 \ m)^2}$$

$F = 5.10$ x 10^5 N □

6.13   **PROBLEM:** When a falling meteor is at a distance above the Earth's surface of 3 times the Earth's radius, what is its acceleration due to the Earth's gravity?

SOLUTION: The acceleration of gravity on Earth $g = Gm/r^2$ follows an inverse-square law. At the surface, at distance one Earth-radius R_e from the center, it is 9.80 m/s^2. At altitude $3R_e$ above the surface, at distance $4R_e$ from the center, the acceleration of gravity will be 4^2 = 16 times smaller:

$$g = \frac{GM_e}{(4R_e)^2} = \frac{GM_e}{16R_e^2} = \frac{9.80 \ m \ / \ s^2}{16} = 0.612 \ m/s^2 \ down \ □$$

6.23 **PROBLEM:** The nucleus of a hydrogen atom, a proton, sets up and electric field. The average distance between the proton and the electron of a hydrogen atom is about 5.1×10^{-11} m. What is the magnitude of the electric field at this distance from the proton?

SOLUTION:

$$\mathbf{E} = \frac{kq}{r^2}\,\hat{\mathbf{r}} = \frac{\left(8.99 \times 10^9 \text{ N} \cdot \text{m}^2\right)\left(1.6 \times 10^{-19}\text{C}\right)}{\left(c^2\right)\left(5.1 \times 10^{-11} \text{ m}\right)^2} \quad \text{away from proton}$$

$\mathbf{E} = 5.53 \times 10^{11}$ N/C away from proton $\square$

6.25 **PROBLEM:** The electron gun in a television tube is to accelerate electrons from rest to 3.0×10^7 m/s within a distance of 2.0 cm. What electric field is required?

SOLUTION: The final speed is related to the acceleration by

$$v^2 - v_0^2 = 2a(x - x_0)$$

$$a = \frac{v^2 - v_0^2}{2(x - x_0)} = 2.25 \times 10^{16} \text{ m/s}^2$$

The magnitude of the force which the electric field exerts on each electron is $|q|E = ma = 2.05 \times 10^{-14}$ N, so the magnitude of the electric field is

$$E = \frac{ma}{|q|} = \frac{m}{|q|}\frac{\left(v^2 - v_0^2\right)}{2(x - x_0)} = \frac{\left(9.11 \times 10^{-31} \text{ kg}\right)\left(3 \times 10^7 \text{ m / s}\right)^2}{\left(1.6 \times 10^{-19} \text{ c}\right)(2)\left(2 \times 10^{-2} \text{ m}\right)}$$

$E = 1.28 \times 10^5$ N/C

The direction of the electric field must be opposite to the direction of the velocity, to exert a forward force on the negative electrons:

$\mathbf{E} = 128$ kN/C opposite to the velocity $\square$

6.29   **PROBLEM:**   A proton moving at 3 x 10^4 m/s is projected at an angle of 30° above a horizontal plane.   If an electric field of 400 N/C is acting downward, how long does it take the proton to return to the horizontal plane?   (*Hint:* Ignore gravity.)

SOLUTION: The proton will follow a path like a ball thrown upward at 30°.   Its upward motion slows and it comes back down because of the electric force on it.

$\mathbf{F_e} = q\mathbf{E} = $ (1.6 x 10^{-19} C) (400 N/C) $(-\mathbf{j})$ = 6.4 x 10^{-17} N$(-\mathbf{j})$,

giving it acceleration

$$\mathbf{a} = \frac{\mathbf{F}}{m} = \frac{6.4 \text{ x } 10^{-17} \text{ N}(-\mathbf{j})}{1.67 \text{ x } 10^{-27} \text{ kg}} = 3.83 \text{ x } 10^{10} \text{ m/s}^2 (-\mathbf{j})$$

The original vertical component of velocity is

(3 x 10^4 m/s) sin 30° = 1.5 x 10^4 m/s.   We are asked to find the time required to move from y_0 = 0 to y = 0, in

$$y = y_0 + v_{y0}t + \frac{1}{2}a_y t^2$$

$$0 = 0 + (1.5 \text{ x } 10^4 \text{ m/s})t + \frac{1}{2}\left(-3.83 \text{ x } 10^{10} \text{ m / s}^2\right)t^2$$

The root t = 0 represents when the proton starts out.   It passes through zero on the way down at time t,

$$1.5 \text{ x } 10^4 \text{ m/s} = \frac{1}{2}\left(+3.83 \text{ x } 10^{10} \text{ m/s}^2\right)t$$

t = 7.83 x 10^{-7} s □

6.33   **PROBLEM:**   A proton is moving at right angles to a magnetic field of 2 T.   What speed does the proton have if the magnetic force on it has a magnitude of 6 x 10^{11} N?

SOLUTION: $\mathbf{F} = q\mathbf{v} \times \mathbf{B}$ or, $\left|\mathbf{F}_m\right| = qvB \sin\theta$ so we have

$$v = \frac{\left(\mathbf{F}_m\right)}{qB\sin\theta} = \frac{6 \text{ x } 10^{-11} \text{ N C} \cdot \text{m}}{\left(1.6 \text{ x } 10^{-19} \text{ C}\right)\left(2 \text{ N} \cdot \text{s}\right)\sin 90°} = 1.88 \text{ x } 10^8 \text{ m/s} □$$

6.39   **PROBLEM:**   Indicate the initial direction of the deflection of charged particles as they enter the magnetic fields, as shown in Figure 6.17.

SOLUTION:

(a)   **v** x **B** is → x ⊗ = = ↑ □

(b)   **v** x **B** is ← x ↑ = = ⊗

= away from you,

so since the charge is negative, q**v** x **B** is (- ⊗) = toward you = ⊙. □

(c)   **v** x **B** is zero since the angle between **v** and **B** is 180° and sin 180° = 0.   There is no deflection. □

(d) v x B is ↑ x ⤢ = = ⊗ = away from you □

6.45 **PROBLEM:** A beam of protons (all with velocity **v**) emerges from a particle accelerator and is deflected in a circular arc with a radius of 0.45 m by a transverse uniform magnetic field of magnitude 0.80 T. (a) Determine the speed v of the protons in the beam. (b) What time is required for the deflection of a particular proton through an angle of 90°?

SOLUTION: The magnetic force is the centripetal force:

$$\Sigma F = ma \qquad qvB \sin \theta = mv^2/r$$

$$v = \frac{qB \sin 90° \, r}{m} = \frac{(1.6 \times 10^{-19} \text{ C})(0.8 \text{ N} \cdot \text{s})(0.45 \text{ m})}{1.67 \times 10^{-27} \text{ kg C} \cdot \text{m}}$$

$$v = 3.45 \times 10^7 \text{ m/s} \ \square$$

Moving with this constant speed, the protons have constant angular speed

$$\omega = \frac{\theta}{t} = \frac{v}{r} = \frac{3.45 \times 10^7 \text{ m / s}}{0.45 \text{ m}} = 7.66 \times 10^7 \text{ rad/s}$$

so to round $\theta = 90° = \pi/2$ takes time

$$t = \frac{\theta}{\omega} = \frac{\pi \text{s}}{2(8.62 \times 10^7 \text{ rad})} = 20.5 \text{ ns} \ \square$$

6.51 **PROBLEM:** Io, a small moon of the giant planet Jupiter, has an orbital period of 1.77 days and an orbital radius of 4.22 x 10^5km. From these data, and using the value G, determine the mass of Jupiter.

SOLUTION: The gravitational force of Jupiter on Io is the centripetal force on Io:

$$\Sigma F_{onIo} = M_{Io}a$$

$$\frac{GM_J M_{Io}}{r^2} = \frac{M_{Io}v^2}{r} = \frac{M_{Io}}{r}\left(\frac{2pr}{T}\right)^2 = \frac{4\pi r M_{Io}}{T^2}$$

Thus

$$M_J = \frac{4\pi^2 r^3}{r} = \frac{4\pi^2\left(4.22 \times 10^8 \text{ m}\right)^3 \text{ kg}^2}{\left(6.67 \times 10^{-11}\text{N} \cdot \text{m}^2\right)\left(1.77 \text{ d}\right)^2}\left(\frac{1 \text{ d}}{86,400 \text{ s}}\right)^2\left(\frac{\text{N} \cdot \text{s}}{\text{kg} \cdot \text{m}}\right)$$

and $M_J = 1.90 \times 10^{27}$ kg $\square$

CHAPTER 7

QUESTIONS

1   **QUESTION:** If the speed of a particle is doubled, what happens to its kinetic energy?

    ANSWER: The object's velocity is tangent to the circle. Its incremental displacement in each tiny time interval is at 90° to the radially inward centripetal force. Now cos 90° = 0 so **F** · d**s** = 0. So as it turns through any angle or through many revolutions, the total work which the centripetal force does is always zero. The centripetal force has no tendency to increase or decrease the object's speed.

7   **QUESTION:** If the speed of a particle is doubled, what happens to its kinetic energy?

    ANSWER: In $K = \dfrac{1}{2}mv^2$, the kinetic energy is proportional to the square of speed, so it gets four times larger when v gets two times larger.

16   **QUESTION:** A team of furniture movers wishes to load a truck
     using a ramp from the ground to the rear of the truck. One of
     the movers claims that less work would be required to load the
     truck if the length of the ramp were increased, reducing the
     angle of the ramp with respect to the horizontral. Is his claim
     valid? Explain.

     ANSWER: Less force will be necessary with a longer ramp, but the
     force must act over a larger distance, to do the same amount of
     work. Suppose we make the frictional force on a refrigerator
     negligible with a wheeled dolly and we roll it up the ramp at
     constant speed. The normal force, at 90° to the motion does no
     work, so the work-energy theorem gives:

     $W_{net} = K_f - K_i$

     $W_{by\ movers} + W_{by\ gravity} = 0$

     The work by gravity adds up just to the weight of the
     refrigerator times the vertical height through which it is
     displaced times (cos 180°). Therefore the movers, however long
     the ramp, must do work *mgh* on the refrigerator.

19   **QUESTION:** A catcher "gives" with the ball when she catches a 0.15-kg baseball moving at 40 m/s. If she moves her catcher's mitt through a distance of 2 cm, what is the average force acting on her hand? $F_{hand\ on\ ball}$

ANSWER: In applying the work-energy theorem, choose initial and final points at the beginning and end of the 2 cm stopping process. Then:

$K_i + W_{net} = K_f$

$$\frac{1}{2}mv_i^2 + F_{hand\ on\ ball}(s)\cos 80°$$

$$F_{hand\ on\ ball} = \frac{mv_i^2}{2s} = \frac{(0.15\ kg)(40\ m\ /\ s)^2}{2\ x\ 0.02\ m} = 6000\ N$$

The force of ball on hand is the same in magnitude but opposite in direction: 6000 N forward. The weight of the ball is negligible in comparison.

CHAPTER 7

PROBLEMS

7.5 **PROBLEM:**   A shopper in a supermarket pushes a cart with a force of 35 N directed at an angle 25° downward from the horizontal. Find the work done by the shopper as she moves down a 50-m length of aisle.

SOLUTION: $W = \mathbf{F} \cdot \mathbf{s} = |\mathbf{F}||\mathbf{s}|\cos\theta = (35 \text{ N})(50 \text{ m})\cos 25° = 1.59 \text{ kJ}$ □

7.11 **PROBLEM:**   For the three vectors $\mathbf{A} = 3\mathbf{i} + \mathbf{j} - \mathbf{k}$, $\mathbf{B} = -\mathbf{i} + 2\mathbf{j} + 5\mathbf{k}$, and $\mathbf{C} = 2\mathbf{j} - 3\mathbf{k}$, find $\mathbf{C} \times (\mathbf{A} - \mathbf{B})$.

SOLUTION: $\mathbf{A} - \mathbf{B} = (3\mathbf{i} + \mathbf{j} - \mathbf{k}) - (-\mathbf{i} + 2\mathbf{j} + 5\mathbf{k}) = 4\mathbf{i} - \mathbf{j} - 6\mathbf{k}$, so
$\mathbf{C} \cdot (\mathbf{A} - \mathbf{B}) = (0\mathbf{i} + 2\mathbf{j} - 3\mathbf{k}) \cdot (4\mathbf{i} - \mathbf{j} - 6\mathbf{k}) = 0(4) + 2(-1) - 3(-6)$
$= 16.0$ □

7.15 **PROBLEM:**   Find the angle between the two vectors given by $\mathbf{A} = -5\mathbf{i} - 3\mathbf{j} + 2\mathbf{k}$ and $\mathbf{B} = -2\mathbf{j} - 2\mathbf{k}$.

SOLUTION: Their dot product is
$\mathbf{A} \cdot \mathbf{B} = (-5\mathbf{i} - 3\mathbf{j} + 2\mathbf{k}) \cdot (0\mathbf{i} - 2\mathbf{j} - 2\mathbf{k})$
$\therefore \mathbf{A} \cdot \mathbf{B} = -5(0) - 3(-2) + 2(-2) = 2$
We also have $|\mathbf{A}| = \sqrt{5^2 + 3^2 + 2^2} = 6.16$
and $|\mathbf{B}| = \sqrt{2^2 + 2^2} = 2.8$
So in $\mathbf{A} \cdot \mathbf{B} = |\mathbf{A}||\mathbf{B}|\cos\theta$
$\theta = \text{Arccos}\left(\dfrac{\mathbf{A} \cdot \mathbf{B}}{AB}\right) = \text{Arccos}\left(\dfrac{2}{6.16 \times 2.83}\right)$
$\theta = \text{Arccos}(0.115) = 83.4°$ □

7.19 PROBLEM: The force acting on a particle varies as in Figure 7.19. Find the work done by the force as the particle moves (a) from $x = 0$ to $x = 8$ m, (b) from $x = 8$ m to $x = 10$ m, and (c) from $x = 0$ to $x = 10$ m.

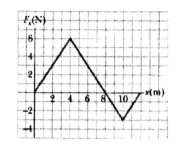

SOLUTION: We identify the work as the area

Figure 7.19

between the x-axis and the F-versus-x graph line

(a) $W_a = F_{avg}\ \Delta x = \left(\dfrac{0 + 6\ \text{N}}{2}\right)(8\ \text{m} - 0) = 24.0$ J ☐

(b) $W_b = F_{avg}\ \Delta x = \left(\dfrac{0 - 3\ \text{N}}{2}\right)(10\ \text{m} - 8\ \text{m}) = -3.00$ J ☐

(c) $W_c = W_a + W_b = 24$ J $-$ 3 J $= 21.0$ J ☐

7.25 PROBLEM: The force required to stretch a Hooke's-law spring varies from zero to 50 N as one end of the spring is moved 12 cm from its unstressed position. (a) Find the force constant k of the spring. (b) Find the work done in stretching the spring.

SOLUTION: We could solve the problem by drawing a graph of applied force versus extension. It would look like Figure 7.10, but with different numbers. The slope is the value of k and the area under the line out to $x = 12$ cm is the work. Alternatively, we can reason from equations: The force exerted by the spring $F_s = -kx$ is equal and opposite to the applied force $F_a = +kx$.

(a) At the maximum extension, 50 N $= K(0.12$ m$)$, so $K = 417$ N/m ☐

(b) $W_{\text{by } F \text{ applied}} = \dfrac{1}{2}kx_m^2 = \dfrac{1}{2}\left(417\ \dfrac{\text{N}}{\text{m}}\right)(0.12\ \text{m})^2 = 3.00$ J ☐

7.27 PROBLEM: A 0.6-kg particle has a speed of 2 m/s at point **A** and kinetic energy of 7.5 J at point **B**.   What are (a) its kinetic energy at point **A**, (b) its speed at point **B**?, and (c) the total work done on the particle as it moves from **A** to **B**?

SOLUTION: (a) $K_A = \frac{1}{2} mv_A^2 = \frac{1}{2}(0.6 \text{ kg})(2 \text{ m/s})^2 = 1.20 \text{ J}$ $\square$

(b) $K_B = \frac{1}{2} mv_B^2;$ $v_B = \sqrt{\frac{2K_B}{m}} = \sqrt{\frac{2(7.5 \text{ J})}{0.6 \text{ kg}} \frac{1 \text{ kg} \cdot \text{m}^2}{1 \text{ J} \cdot \text{s}^2}} = 5.00 \text{ m/s}$ $\square$

(c) $K_A + W_{net} = K_B$

$1.20 \text{ J} + W_{net} = 7.50 \text{ J}$ $W_{net} = \Delta K = 6.30 \text{ J}$ $\square$

7.31 PROBLEM: A 3-kg mass has an initial velocity $v_0 = (6i - 2j)$ m/s. (a) What is its kinetic energy at this time?   (b) Find the *change* in its kinetic energy if its velocity changes to (8i + 4j) m/s. (*Hint:* Remember that $v^2 = v \cdot v$.)

SOLUTION:

(a) $K_0 = \frac{1}{2} mv_0^2 = \frac{1}{2} m \, v_0 \cdot v_0 = \frac{1}{2}(3 \text{ kg})(6i - 2j)\text{m/s} \cdot (6i - 2j)\text{m/s}$

$= (1.5 \text{ kg})(36 + 4)\text{m}^2/\text{s}^2 = 60.0 \text{ J}$ $\square$

(b) $K_f = \frac{1}{2} m v_f \cdot v_f = \frac{1}{2}(3 \text{ kg})(8i + 4j)\text{m/s} \cdot (8i + 4j)\text{m/s}$

$= (1.5 \text{ kg})(64 + 16) \text{ m}^2/\text{s}^2 = 120 \text{ J}$

$W_{net} = DK = K_f - K_i = 120 \text{ J} - 60 \text{ J} = 60.0 \text{ J}$ $\square$

It would be an extra unnecessary step to work out the original and final speeds, $\sqrt{40}$ m/s and $\sqrt{80}$ m/s.

7.35 **PROBLEM:** A 15-kg block is dragged over a horizontal surface by a constant force of 70 N acting at an angle of 20° above the horizontal. The block is displaced 5 m, and the coefficient of kinetic friction between the block and surface is 0.3. Find the work done by (a) 70-N force, (b) the normal force, and (c) the force of gravity. (d) Calculate the energy lost due to friction.

SOLUTION: (a) $w = \mathbf{F} \cdot \mathbf{s} = (70 \text{ N})(5 \text{ m})\cos 20° = 329 \text{ J}$ $\square$

(b) The normal force is perpendicular to the motion, so its dot product with the displacement is zero $\square$.

(c) The motion is horizontal and the weight is vertical so $W = 0$ since $\cos 90° = 0$

(d) We must find the force of friction by analyzing the vertical forces:

$\Sigma F_y = ma_y$

$(70 \text{ N})\sin 20° + n - (15 \text{ kg})(9.8 \text{m/s}^2) = 0$

$n = 147 \text{ N} - 23.9 \text{ N} = 123 \text{ N}$

Now $f_K = \mu_K n = 0.3 \times 123 \text{ N} = 36.9 \text{ N}$

and $\Delta K_{friction} = -fs = -(36.9 \text{ N})(5 \text{ m}) = -185 \text{ J}$ $\square$

The total work on the object is

$329 \text{ J} + 0 + 0 - 185 \text{ J} = 144 \text{ J}$.

This will be the change in its kinetic energy.

7.39 **PROBLEM:** A sled of mass m on a frozen pond is given a kick, which imparts to it an initial speed $v_0 = 2$ m/s. The coefficient of kinetic friction between the sled and the ice is $\mu_k = 0.1$. Find the distance the sled moves before coming to rest.

SOLUTION: Step one: analyze the vertical forces,

$$\Sigma F_y = ma_y$$
$$+n - mg = 0$$
$$n = mg$$

Step two: identify the size of the friction force:

$$f_K = m_K\, mg$$

Step three: Apply the work-energy theorem. We do not consider the kick, but take the original point after the kick, with $v_0 = 2$ m/s, and the final point where the sled stops moving. The weight and normal force, at 90° to the motion, do no work.

$$K_i + W_{net} = K_F$$

$$\frac{1}{2}mv_0^2 - fs = 0$$

$$s = \frac{mv_0^2}{2f} = \frac{mv_0^2}{2\,\mu_K\, mg} = \frac{v_0^2}{2\,\mu_k\, g} = \frac{(2 \text{ m/s})^2}{2(0.1)(9.8 \text{ m/s}^2)}$$

$$s = 2.04 \text{ m } \square$$

7.41 **PROBLEM:** A 700-N marine in basic training climbs a 10-m vertical rope, at uniform speed, in 8 s. What is his effective power output?

SOLUTION: The marine must exert a 700 N upward force to lift his body at constant speed. Then his muscles do work
$W = \mathbf{F} \cdot \mathbf{s} = 700$ N up $\cdot$ 10 m up $= 7000$ J. The power he puts out is

$$P = \frac{W}{t} = \frac{7000 \text{ J}}{8 \text{ s}} = 875 \text{ W } \square$$

7.44   **PROBLEM:**   A 1500-kg car accelerates uniformly from rest to a
speed of 10 m/s in 3 s.   Neglecting friction between car and
highway and within the car, find (a) the work done on the car in
this time, (b) the average power delivered by the engine in the
first 3 s, and (c) the instantaneous power delivered by the
engine at $t = 2s$.

SOLUTION: We imagine there is no kinetic friction between car
and highway, and no air resistance.   But there has to be static
friction between tires and road to push the car forward; otherwise the car would remain in one place with its wheels spinning.
To oppose the bottom of the tire sliding backward on the road,
static friction pushes forward on the tire and does positive
work on the car body:

$$K_i + W_{net} = K_f$$

$$0 + W_{by\ friction} = \frac{1}{2}mv_f^2 = \frac{1}{2}(1500\ kg)(10\ m/s)^2 = 75,000\ J\ \square$$

(b) $P = \dfrac{W}{t} = \dfrac{75,000\ J}{3\ s} = 25,000\ W\ \square$

(c) The car moves distance $x = \frac{1}{2}(v_0 + v_f)t$, or

$$x = \frac{1}{2}mv_f^2 = \frac{1}{2}(1500\ kg)(10\ m/s)^2 = 15\ m,\ \text{with acceleration}$$

$$a = \frac{v_f - v_0}{t} = \frac{10\ m/s - 0}{3\ s} = 3.33\ m/s^2.$$

So its speed at $t = 2s$ is
$v = v_0 + at = 0 + 3.33\ m/s^2(2s) = 6.67\ m/s$.   We can see how
large the propelling friction force is from
$W_{by\ friction} = fs\ \cos 0° = 75,000\ J$

$$f = \frac{75,000\ J}{15\ m} = 5000\ N.$$

Then the instantaneous power at $t = 2$ s is $P = F \cdot v$

$P = 5000\ N$ forward $\cdot\ 6.67$ m/s forward $= 33.3$ kW $\square$

The force and acceleration are constant in time, but the
power must increase as the speed increases.

7.57 PROBLEM: A pile driver of mass 2100 kg is used to drive a steel I-beam into the ground. The mass falls freely from rest a distance of 5 m before contacting the beam, and it drives the beam 12 cm into the ground before coming to rest. Using the work-energy relation, calculate the average force that the beam exerts on the mass while the mass is brought to rest.

SOLUTION: Choose the initial point when the mass is elevated and the final point when it comes to rest again 5.12 m below. Two forces do work on it: gravity and the normal force of the beam:

$K_i + W_{net} = K_f$

$0 + Ws_w\cos 0° + ns_n\cos 180° = 0$

Notice here that the weight pulls in the direction of motion and the beam pushes upward, which is backward. Next

$(2100 \text{ kg})(9.8 \text{ m/s}^2)(5.12 \text{ m}) + n(0.12 \text{ m})(-1) = 0$

$$n = \frac{1.05 \times 10^5 \text{ J}}{0.12 \text{ m}} = 878 \text{ kN up } \square$$

In the next chapter we will see another way to describe the process:

The 2100-kg mass falls 5.12 m and its energy mgh does work $F \cdot s$

$mgh = F \cdot s$

$(2100)(9.8)(5.12) = F(0.12)$

and $F = 878,000 \text{ N}$

is the force exerted on the beam by the 2100-kg mass.

7.59 PROBLEM: A 200-g block is pressed against a spring of force constant 1400 N/m until the block compresses the spring 10 cm. The spring rests at the bottom of a ramp incline at 60° to the horizontal. Determine how far up the incline the block moves before momentarily coming to rest, (a) if there is no friction between the block and the ramp and (b) if the coefficient of kinetic friction is 0.4.

7.59 (cont.)

SOLUTION: Let d represent the distance that the block slides along the incline. Then the work done by gravity is $W = \mathbf{F} \cdot \mathbf{s}$ = $mg\, d \cos 150°$. Note how the angle is measured between the downward force and the displacement up the incline. The normal force does zero work, being at right angles to the displacement. From Equation 7.14, the work done by the spring force on the block is $\frac{1}{2}kx_m^2$. We apply the work-energy theorem between the starting-point of the motion and the maximum excusion up the incline:

$$K_i + W_{net} = K_f$$

$$0 + mgd \cos 150° + 0 + \frac{1}{2}kx_m^2 = 0$$

$$d = \frac{-kx_m^2}{2mg \cos 150°} = \frac{(-1400 \text{ N/m})(0.1 \text{ m})^2}{2(0.2 \text{ kg})(9.8 \text{ m/s}^2)(-0.866)}$$

$$d = \frac{-kx_m^2}{2mg \cos 150} = \frac{(-140 \text{ N/m})(0.1 \text{ m})^2}{2(0.2 \text{ kg})(9.8 \text{ m /s}^2)(-0.866)}$$

$$d = \frac{7 \text{ J}}{(1.96 \text{ N})(0.866)} = 4.12 \text{ m} \quad \square$$

(b) The analysis of forces perpendicular to the incline reads

$$\Sigma F_y = ma_y \qquad +n - (1.96 \text{ N})\cos 60° = 0 \qquad n = 0.98 \text{ N}$$

Then the force of friction is

$$f_k = \mu_k n = 0.4 \times 0.98 \text{ N} = 0.392 \text{ N}$$

Now the work-energy theorem contains one more term:

$$K_i + W_{net} = K_f$$

$$0 + mgd \cos 150° + f_k d \cos 180° + \frac{1}{2}kx_m^2 = 0$$

$$(1.96 \text{ N})d(-0.866) + (0.392 \text{ N})d(-1) + 7 \text{ J} = 0$$

$$d = \frac{7 \text{ J}}{1.70 \text{ N} + 0.392 \text{ N}} = 3.35 \text{ m} \quad \square$$

CHAPTER 8

QUESTIONS

4   **QUESTION:** Discuss the energy transformations that occur during a pole vault event. Ignore rotational motion.

ANSWER: Chemical energy in the athlete's body is converted mostly into heat but also into kinetic energy as he runs. Now with little lost heat the kinetic energy becomes elastic potential energy in the bent pole, then gravitational energy in the elevated body of the vaulter, and then kinetic energy as he falls. When he lands on the pad, this energy immediately and totally turns into extra thermal energy, mostly in the pad.

17   **QUESTION:** Three identical balls are thrown from the top of a building, all with the same initial speed. One ball is thrown horizontally, the second at some angle above the horizontal, and the third at some angle below the horizontal. Neglecting air resistance, compare the speeds of the balls as they reach the ground.

ANSWER: The three balls have different patterns of change in speed. The ball thrown upward first slows down, while the other two balls always speed up. The three take different times to reach the ground. But all have the same impact speed, since all start with the same kinetic energy and undergo conversion of the same amount of gravitational into extra kinetic energy.
($E_{total} = mgh + \frac{1}{2}mv^2$ for all three balls.)

20   **QUESTION:**   Give a physical explanation of the fact that the potential energy of a pair of like charges is positive, whereas the potential energy of a pair of unlike charges is negative.

ANSWER: When they are close together, two like charges have higher potential energy than when they are far apart - you can tell because they gain kinetic energy in springing apart.   Two opposite charges have lower potential energy when they are close together than when far apart, since you must put in work to pull them apart at constant speed.   We arbitrarily choose zero potential energy to describe infinite separation (where the force is zero and the particles can be said to be too far apart to interact).   So we must take the potential energy of like charges as greater than zero when they are close together, and the potential energy of opposite charges as negative.

CHAPTER 8

PROBLEMS

8.5   **PROBLEM:**   A 3-kg particle moves from the
origin to the position having coordinates
x = 5 m and y = 5 m, under the influence of
gravity acting in the negative y direction
(Figure 8.5).   Using Equation 8.1, calculate
the work done by gravity when the particle goes from O to C
along the following paths: (a) OAC, (b) OBC, (c) OC.   Your
results should Figure 8.5 all be identical.   Why?

SOLUTION:

(a) $W_{\text{by gravity}} = F_{\text{gravity}} s_{OA} \cos \theta \; + \; F_g s_{AC} \cos \theta$

$= (3 \text{ kg})(9.8 \text{ m/s}^2)(5 \text{ m}) \cos 90° + (29.4 \text{ N})(5 \text{ m}) \cos 180°$

$= -147 \text{ J} \; \square$

(b) $W_g = \mathbf{F_g \cdot s_{OB}} + \mathbf{F_g \cdot s_{BC}}$

$= 29.4 \text{ N}(-\mathbf{j}) \cdot (5 \text{ m})\mathbf{j} + 29.4 \text{ N}(-\mathbf{j}) \cdot (5 \text{ m})\mathbf{i}$

$= -147 \text{ J} + 0 = -147 \text{ J} \; \square$

(c) $W_g = (29.4 \text{ N})\sqrt{(5 \text{ m})^2 + (5 \text{ m})^2} \cos(90° + 45°)$

$= (29.4 \text{ N})(7.07 \text{ N}) \cos 135° = -147 \text{ J} \; \square$

The results are identical because gravity is a conservative
force.   The work it does is path-independent and always equal to
the negative of the change in gravitational potential energy.

8.13 **PROBLEM:** A rocket is launched, at an angle of 53° to the horizontal, from altitude h with speed v_0. (a) Use energy methods to find the speed of the rocket when its altitude is $h/2$. (b) Find the x and y components of velocity when the rocket's altitude is $h/2$, using the fact that $v_x = v_{x0}$ is constant (since $a_x = 0$) and the result of part (a).

SOLUTION: We assume the rocket engine is turned off immediately after launch, so the rocket moves as a projectile and energy is conserved. Choose the point just after launch as the original point and the point at altitude $h/2$ as the final point. Then

$$(K + U_g)_0 = (K + U_g)_f$$

gives $$\frac{1}{2}mv_0^2 + mgh = \frac{1}{2}mv_f^2 + mg\frac{h}{2}$$

$$v_0^2 + 2gh - gh = v_f^2$$

$$v_f = \sqrt{v_0^2 + gh} \quad \square$$

(b) The x-component of velocity is always $v_{x0} = v_0\cos 53° \; \square$, so the y-component at this point must be, from

$$v_f^2 = v_{xf}^2 + v_{yf}^2$$

$$v_{yf} = \sqrt{v_f^2 - v_{xf}^2} = -\sqrt{v_0^2 + gh - v_0^2\cos^2 53°}$$

$$v_{yf} = -\sqrt{v_0^2(1 - \cos^2 53°) + gh} = -\sqrt{v_0^2\sin^2 53° + gh} \quad \square$$

We choose the negative root because the object must be headed down.

8.17   **PROBLEM:**   Two masses are connected by a light
        string that passes over a light, frictionless
        pulley as in Figure 8.17.   The 5-kg mass is
        released from rest.   Using the law of
        conservation of energy, (a) determine the
        velocity of the 3-kg mass just as the 5-kg mass
        hits the ground, and (b) find the maximum
        height to which the 3-kg mass will rise.

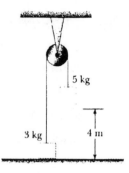

Figure 8.17

SOLUTION: The tension in the cord provides the force that
allows the heavier mass to lift the lighter one, but the total
energy of both objects together stays constant.   The work-
energy theorem for the two objects A and B is $(K_A + K_B + U_A +$
$U_B)_i + \Delta K_{nc} = (K_A + K_B + U_A + U_B)_f$
Choose the initial point before release and the final point just
before the larger mass hits the floor:

$$0 + 0 + (5 \text{ kg})(9.8 \text{ m/s}^2)(4 \text{ m}) + 0 + 0$$
$$= \tfrac{1}{2}(5 \text{ kg})v_f^2 + \tfrac{1}{2}(3 \text{ kg})v_f^2 + 0 + (3 \text{ kg})(9.8 \text{ m/s}^2)(4 \text{ m})$$
$$(2 \text{ kg})(9.8 \text{ m/s}^2)(4 \text{ m}) = (8 \text{ kg})v_f^2$$
$$v_f = 4.43 \text{ m/s} \ \square$$

(b) Now the string goes slack.   The 5-kg mass loses all its
    mechanical energy, but the 3-kg mass becomes a projectile.
    Take the initial point at the previous final point, and the
    new final point at its maximum height:

$$(K + U_g)_i = (K + U_g)_f$$
$$\tfrac{1}{2}(3 \text{ kg})(4.43 \text{ m/s})^2 + (3 \text{ kg})(9.8 \text{ m/s}^2)(4 \text{ m})$$
$$= 0 + (3 \text{ kg})(9.8 \text{ m/s}^2)y_f$$

$y_f = 5.00 \text{ m} \ \square$ or 1 m higher than the height of the
5-kg mass when it was released.

8.19 PROBLEM: A 5-kg block is set in motion up an inclined plane, as in Figure 8.19, with an initial speed of 8 m/s. The block comes to rest after traveling 3 m along the plane, as shown in the diagram. The plane is inclined at an angle of 30° to the horizontal. (a) Determine the change in kinetic energy. (b) Determine the change in potential energy. (c) Determine the frictional force on the block (assumed to be constant). (d) What is the coefficient of kinetic friction?

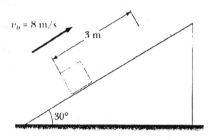

Figure 8.19

SOLUTION:

(a) $\Delta K = K_f - K_i = 0 - \frac{1}{2}(5 \text{ kg})(8 \text{ m/s})^2 = -160 \text{ J}$ ☐

(b) $\Delta U = U_{gf} - U_{gi} = mgy_f - 0 = (5 \text{ kg})(9.8 \text{ m/s}^2)(3 \text{ m})\sin 30°$

   $\Delta U = 73.5 \text{ J}$ ☐

(c) $(K + U_g)_i + \Delta K_{nc} = (K + U_g)_f$

   $\frac{1}{2}(5 \text{ kg})(8 \text{ m/s})^2 + 0 + f(3 \text{ m})\cos 180°$

   $\qquad = 0 + (5 \text{ kg})(9.8 \text{ m/s}^2)(1.5 \text{ m})$

   $160 \text{ J} - f(3 \text{ m}) = 73.5 \text{ J}$

   $f = \dfrac{86.5 \text{ J}}{3 \text{ m}} = 28.8 \text{ N}$ ☐

(d) The forces perpendicular to the incline must add to zero:

   $\Sigma F_y = 0$

   $+n - mg \cos 30° = 0$

   $n = (5 \text{ kg})(9.8 \text{ m/s}^2)\cos 30° = 42.4 \text{ N}$

   Now $f_k = \mu_k n$ gives

   $\mu_k = \dfrac{f_k}{n} = \dfrac{28.8 \text{ N}}{42.4 \text{ N}} = 0.679$ ☐

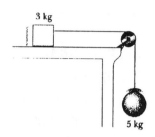

Figure 8.27

8.27 **PROBLEM:** The coefficient of friction between the 3.0-kg object and the surface in Figure 8.27 is 0.40. The masses start from rest. What is the speed of the 5.0-kg mass when it has fallen a vertical distance of 1.5 m?

SOLUTION: We could solve this problem by using $\Sigma F = ma$ to give a pair of simultaneous equations in the unknowns acceleration and tension; then we would have to solve a motion problem to find final speed. It is easier to solve from the work-energy theorem. We begin the same way, by saying for the 3-kg block

$w = mg = (3 \text{ kg})(9.8 \text{ m/s}^2) = 29.4 \text{ N}$

$\Sigma F_y = ma_y$

$+n - 29.4 \text{ N} = 0$

$n = 29.4 \text{ N}$

$f_k = \mu_k n = 0.40 \times 29.4 \text{ N} = 11.8 \text{ N}$

Now for two objects A and B the work-energy theorem is

$(K_A + K_B + U_A + U_B)_i + \Delta K_{nc} = (K_A + K_B + U_A + U_B)_f$

Choose the initial point before release and the final point after each block has moved 1.5 m. For the 5-kg block the zero level of gravitational energy is at the final position. For the 3-kg block choose $U_g = 0$ at the tabletop.

8.27 (cont.)

Then we have

$$0 + 0 + 0 + m_B g y_{Bi} + fs \cos \theta = \frac{1}{2} m_A v_f^2 + \frac{1}{2} m_B v_f^2 + 0 + 0$$

$$(5 \text{ kg})(9.8 \text{ m/s}^2)(1.5 \text{ m}) + (11.8 \text{ N})(1.5 \text{ m})\cos 180°$$

$$= \frac{1}{2}\left(3 \text{ kg}\right) v_f^2 + \frac{1}{2}\left(5 \text{ kg}\right) v_f^2$$

$$73.5 \text{ J} - 17.6 \text{ J} = \frac{1}{2}\left(8 \text{ kg}\right) v_f^2 = \sqrt{\frac{2 \times 55.9 \text{ J}}{8 \text{ kg}}} = 3.74 \text{ m/s} \ \square$$

Yet another solution is to observe the total energy given to motion and friction comes from the 5 kg potential energy:

$$Mgh = \mathbf{f} \cdot \mathbf{s} + \tfrac{1}{2}mv^2 + \tfrac{1}{2}Mv^2$$

Solving, $v^2 =$

$$h = s = 1.5 \text{ m} \qquad M = 5 \text{ kg}$$

$$m = 3 \text{ kg} \qquad \mu = 0.4$$

$$v = 3.74 \text{ m/s}$$

8.33 **PROBLEM:** After our Sun exhausts its nuclear fuel, its ultimate fate is possibly to collapse to a *white dwarf* state, in which it has approximately the mass of the Sun but the radius of the Earth. Calculate (a) the average density of the white dwarf, (b) the acceleration due to gravity at its surface, and (c) the gravitational potential energy of a 1-kg object at its surface. (Take U_g = 0 at infinity.)

SOLUTION:

(a) $\rho = \dfrac{M_s}{V} = \dfrac{M_s}{4p\,R_e^2\,/\,3} = \dfrac{1.99 \text{ x } 10^{30} \text{ kg}}{4\pi(6.37 \text{ x } 10^6 \text{ m})^3 / 3}$

$\rho = 1.84 \text{ x } 10^9 \text{ kg/m}^3$ □

(b) For an object of mass m on its surface,

$mg = \dfrac{GM_s M}{R_e^2},$ so

$g = \dfrac{GM_s}{R_e^2} = \dfrac{(6.67 \text{ x } 10^{-11} \text{ N} \cdot \text{m}^2)(1.99 \text{ x } 10^{30} \text{ kg})}{(\text{kg}^2)(6.37 \text{ x } 10^6 \text{ m})^2}$

$g = 3.27 \text{ x } 10^6 \text{m/s}^2$ □

(c) $U_g = \dfrac{-GM_s m}{R^e}$

$U_g = \dfrac{(-6.67 \text{ x } 10^{-11} \text{ N} \cdot \text{m}^2)(1.99 \text{ x } 10^{30} \text{ kg})(1 \text{ kg})}{(\text{kg}^2)(6.37 \text{ x } 10^6 \text{ m})}$

$U_g = -2.08 \text{ x } 10^{13}$ J □

8.39 **PROBLEM:** Calculate the energy required to assemble the array of charges shown in Figure 8.39, where $a = 0.20$ m, $b = 0.40$ m, and $q = 6$ mC.

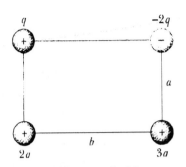

Figure 8.39

SOLUTION: Imagine starting with all charges very far away, so that they exert negligible forces on each other. Zero work is required to bring up the charge q to its final location. Then to bring up the $-2q$ charge requires work to change the potential energy to $U = \dfrac{k_e q_1 q_2}{r} = \dfrac{k_e \, q(-2q)}{b}$ to bring up the $2q$ charge close to the other two requires extra work input $\dfrac{k_e q(2q)}{a} + \dfrac{k_e(-2q)(2q)}{\sqrt{a^2 + b^2}}$

Then to bring up the $3q$ charge requires still more work in amount $\dfrac{k_e q(3q)}{\sqrt{a^2 + b^2}} + \dfrac{k_e(-2q)(3q)}{a} + \dfrac{k_e(2q)(3q)}{b}$

The total energy required to assemble the array is

$$k_e q^2 \left(-\frac{2}{b} + \frac{2}{a} - \frac{4}{\sqrt{a^2 + b^2}} + \frac{3}{\sqrt{a^2 + b^2}} - \frac{6}{a} + \frac{6}{b} \right)$$

$$= k_e q^2 \left(\frac{4}{b} - \frac{4}{a} - \frac{1}{\sqrt{a^2 + b^2}} \right)$$

$$= (8.99 \times 10^9 \text{ N} \cdot \text{m}^2/\text{C}^2)(6 \times 10^{-6} \text{ C})^2 \left(\frac{4}{0.4 \text{ m}} - \frac{4}{0.2 \text{ m}} - \frac{1}{0.447 \text{ m}} \right)$$

$$= -3.96 \text{ J} \quad \square$$

8.45 **PROBLEM:** A particle of mass m = 5 kg is released from point A on the frictionless track shown in Figure 8.45. Determine (a) the speed of mass m at points B and C and (b) the net work done by the force of gravity in moving the particle from A to C.

SOLUTION: (a) There is no friction. The normal force, which is always perpendicular to the motion, does no work. So energy is conserved:

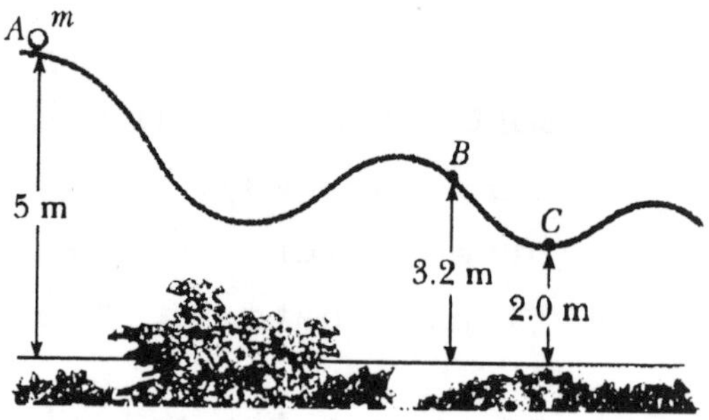

$$(K + U)_A + \Delta K_{nc} = (K + U)_B$$

$$0 + mgy_A + 0 = \tfrac{1}{2} mv + mgy_B$$

The mass makes no difference:

Figure 8.45

$$v_B = \sqrt{2g(y_A - y_B)} = \sqrt{2 \times 9.8 \text{ m/s}^2 \times 1.8 \text{ m}} = 5.94 \text{ m/s} \ \square$$

Again,

$$(K + U)_A = (K + U)_C$$

$$0 + mgy_A = \tfrac{1}{2} mv_C^2 + mgy_C$$

$$v_C = \sqrt{2g(y_A - y_C)} = \sqrt{2 \times 9.8 \text{ m/s}^2 \times 3 \text{ m}} = 7.67 \text{ m/s} \ \square$$

(b) The work done by gravity is the negative of the change in gravitational potential energy:

$$-\Delta U = -(U_C - U_A) = U_A - U_C = mg(y_A + y_C)$$

$$-\Delta U = (5 \text{ kg})(9.8 \text{ m/s}^2)(3 \text{ m}) = 147 \text{ J} \ \square$$

8.47 **PROBLEM:** The potential energy of a two-particle system separated by a distance r is given by $U(r) = A/R$, where A is a constant. Find the radical force $\mathbf{F_r}$.

SOLUTION: The force is the negative derivative of the potential energy with respect to distance:

$$F_r = -\frac{dU}{dr} = -\frac{d}{dr}\left(\frac{A}{r}\right) = -A(-1)\,r^{-2} = \frac{A}{r^2}$$

This describes an inverse-square-law force of repulsion, as between two negative point electric charges.

8.55 **PROBLEM:** A 10-kg block is released from point *A* on track *ABCD* (Figure 8.55). The track is frictionless except for the portion *BC*, which is 6 m long. The block travels down the track, hits a spring of force constant k = 2250 N/m, and compresses it a distance of 0.3 m from its equilibrium position before coming to rest momentarily. Determine the coefficient of kinetic friction between the track portion *BC* and the block.

SOLUTION: Choose the initial point when the block is at *A* and the final point when the spring is fully compressed:

$$(K + U_g + U_s)_A + \Delta K_{nc} = (K + U_g + U_s)_f$$

$$0 + mgy_A + 0 + fs_{BC}\cos 180° = 0 + 0 + \tfrac{1}{2}kx_f^2$$

$$(10 \text{ kg})(9.8 \text{ m/s}^2)(3 \text{ m}) - f(6 \text{ m}) = \tfrac{1}{2}(2250 \text{ N/m})(0.3 \text{ m})^2$$

$$294 \text{ J} - f(6 \text{ m}) = 101 \text{ J}$$

$$f = \frac{193 \text{ J}}{6 \text{ m}} = 32.1 \text{ N}$$

Now about the vertical forces when the block is between *B* and *C*:

$$\Sigma F_y = 0$$

$$+n - mg = 0$$

$$n = (10 \text{ kg})(9.8 \text{ m/s}^2) = 98 \text{ N}$$

So, $f_K = m_K n$

$$\mu_K = \frac{f_K}{n} = \frac{32.1 \text{ N}}{98 \text{ N}} = 0.328 \; \square$$

Another way to look at the problem is to say that the original potential energy mgh goes into frictional energy loss, and into compressing the spring:

$$mgh = \mathbf{f \cdot s} + \tfrac{1}{2}kx^2$$

$$\mathbf{f \cdot s} = \mu mg \cdot s = mgh - \tfrac{1}{2}kx^2$$

m = 10 kg h = 3 m

k = 2250 N/m x = 0.3 m

s = 6.0 m

$$\mu = \frac{h}{s} - \frac{kx^2}{2mgs} = 0.328$$

8.62 **PROBLEM:** Two blocks, A and B (with masses of 50 kg and 100 kg, respectively), are connected by a string as shown in Figure 8.62. The pulley is frictionless and of negligible mass. The coefficient of kinetic friction between block A and the incline is $\mu_k = 0.25$. Determine the change in the kinetic energy of block A as it moves from C to D, a distance of 20 m up the incline.

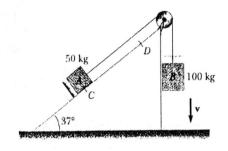

Figure 8.62

SOLUTION: The blocks are shown already moving at point C. We need not know how fast to determine the change in kinetic energy, $K_{AD} - K_{AC}$. We suppose the 100-kg mass does not crash into the floor before the 50-kg mass reaches point D. The weight of Block A, $w = mg = (50 \text{ kg})(9.8 \text{ m/s}^2) = 490 \text{ N}$ has component $(490 \text{ N})\cos 37°$ perpendicularly into the incline, so $\Sigma F_y = ma_y$ gives $+n - (490 \text{ N})\cos 37° = 0$, $n = 392 \text{ N}$ and $f_k = \mu_k n = 0.25 \times 392 \text{ N} = 97.8 \text{ N}$.

Now the work-energy theorem for the two objects is

$$(K_A + K_B + U_A + U_B)_C + \Delta K_{nc} = (K_A + K_B + U_A + U_B)_D$$

For Block A take $y_C = 0$. Then $y_D = (20 \text{ m})\sin 37°$

For Block B take $y_D = 0$. Then $y_C = 20 \text{ m}$, and

$$\tfrac{1}{2} m_A v_C^2 + \tfrac{1}{2} m_B v_C^2 + 0 + m_B g y_C + fs \cos 180°$$

$$= \tfrac{1}{2} m_A v_D^2 + \tfrac{1}{2} m_B v_D^2 + m_A g y_D + 0$$

$$(100 \text{ kg})(9.8 \text{ m/s}^2)(20 \text{ m}) + (97.8 \text{ N})(20 \text{ m})(-1)$$
$$- (50 \text{ kg})(9.8 \text{ m/s}^2)(20 \text{ m}) \sin 37°$$

$$= \tfrac{1}{2}(150 \text{ kg})(v_D^2 - v_C^2) = 11745 \text{ J}$$

This is the change in kinetic energy of both blocks. For block A alone, one-third of the total mass, it is

$$K_{AD} - K_{AC} = \frac{1}{2}(50 \text{ kg})(v_D^2 - v_C^2)$$

$$= \frac{1}{3}\frac{1}{2}(150 \text{ kg})(v_D^2 - v_C^2) = \frac{11745 \text{ J}}{3}$$

$$= 3.92 \text{ kJ} \quad \square$$

8.65 **PROBLEM:** In the dangerous "sport" of bungee-jumping, a daring student jumps from a balloon with a specially designed elastic cord attached to his ankles, as shown in the photograph. The unstretched length of the cord is 25 m, the student weights 700 N, and the balloon is 36 m above the surface of a river below. Calculate the required force constant of the cord if the student is to stop safely 4 m above the river.

SOLUTION: The weight of the student $w = mg = 700$ N. The distance the student moves down is 32 m.

Therefore, the bungee cord must absorb energy $\frac{1}{2}k(\Delta x)^2$ by stretching such that $mgh = \frac{1}{2}k(\Delta x)^2$ where $mg = 700$ N, $h = 32$ m, $\Delta x = 7$ m, and k is unknown. Solving for k,

$$k = \frac{2mgh}{(\Delta x)^2} = \frac{2(700)(32)}{(7)^2} = 914 \text{ N/m} \quad \square$$

8.66 **PROBLEM:** A small object with charge -4 nC and mass 0.3 g moves freely after being released form rest 80 cm from a *fixed* spherical object with charge +5 μC and radius 6 cm. Find the speed of the 0.3-g object just before it collides with the sphere.

SOLUTION: The small object gains kinetic energy and loses electric potential energy, keeping constant total energy as it moves freely from radius 86 cm to radius 6 cm in the field of the sphere of charge. This field is identical to that of a 5 μC point charge at the center of the sphere, so we may also think of this point charge to compute the potential energy:

$$(K + U)_i = (K + U_f)$$

$$0 + k_e \frac{q_1 q_2}{r_i} = \frac{1}{2} m v_f^2 + k_e \frac{q_1 q_2}{r_f}$$

$$\frac{8.99 \times 10^9 \text{ N} \cdot \text{m}^2}{c^2} \frac{(-4 \times 10^{-9} \text{ C})(5 \times 10^{-6} \text{ C})}{0.86 \text{ m}}$$

$$= \frac{1}{2}(3 \times 10^{-4} \text{ kg}) v_f^2 + \frac{8.99 \times 10^9 \text{ N} \cdot \text{m}^2}{c^2} \frac{(-4 \times 10^{-9} \text{ C})(5 \times 10^{-6} \text{ C})}{(0.06 \text{ m})}$$

$$-2.09 \times 10^{-4} \text{ J} = \frac{1}{2}(3 \times 10^{-4} \text{ kg}) v_f^2 - (3.00 \times 10^{-3} \text{ J})$$

$$\sqrt{\frac{2(2.79 \times 10^{-3} \text{ J})}{3 \times 10^{-4} \text{ kg}}} = v_f = 4.31 \text{ m/s} \quad \square$$

CHAPTER 9

QUESTIONS

5   **QUESTION:** Explain how momentum is conserved when a ball bounces
     from a floor.

     ANSWER: The ball's downward momentum increases as it
     accelerates down. A larger momentum change occurs when it
     touches the floor and switches to upward motion. The outside
     forces of gravity and the normal force inject impulses to change
     its momentum. If we think of ball-and-Earth-together as our
     system, these forces are internal and do not change the total
     momentum. It is conserved, if we neglect the curvature of the
     Earth's orbit. As the ball falls down, the Earth lurches up to
     meet it, on the order of 10^{25} times more slowly. Ball and Earth
     bounce off each other and separate. You dribble the Earth as
     you dribble the ball.

12   **QUESTION:** A researcher tranquilized a polar bear on a friction-
     less glacier. How might the researcher, knowing her own weight,
     be able to *estimate* the weight of the polar bear using a
     measuring tape and a rope?

     ANSWER: Tie one end of the rope around the bear. Lay out the
     tape measure on the ice between the bear's original location and
     yours, as you hold the other end of the rope. Take off your
     spiked shoes and pull on the rope. Both you and bear will glide
     over the ice until you meet. From the tape, observe how far the
     bear has slid, l_b, and how far you have, l_y. The point where
     you meet is the constant location of the center of mass of the
     bear-plus-you system, so you can determine the mass of the bear
     from $m_b l_b = m_y l_y$. The bear now wakes up and you cannot get back
     to your spiked shoes.

25 **QUESTION:** An airbag is inflated when a collision occurs, which protects the passenger (the dummy, in this case) from serious injury. Why does the airbag soften the blow? Discuss the physics involved in this dramatic photograph.

ANSWER: Crashing your car into a brick wall, you must come to a stop. Some outside force must inject you with backward impulse equal to your original forward momentum. You may take your choice between getting this impulse as a large force over a short time, exerted by the wall after you have crashed out through the windshield; or a smaller force over a longer time, exerted by the air bag as your body squashes into it. The inflated air bag extends the distance over which your body loses its speed, so it extends your stopping time.

CHAPTER 9

PROBLEMS

9.5 **PROBLEM:** A 1500-kg car moving with a speed of 15 m/s collides with a utility pole and is brought to rest in 0.3 s. Find the average force exerted on the car during the collision.

SOLUTION: $\int F dt = \Delta \mathbf{p}$

$\mathbf{F}_{avg} \Delta t = \mathbf{p}_f - \mathbf{p}_i$

We choose the initial and final points 0.3 s apart. Choose the x-axis in the direction of motion.

$(\mathbf{F})(0.3 \text{ s}) = (1500 \text{ kg})(0) - (1500 \text{ kg})(15 \text{ m/s})\mathbf{i}$

$\mathbf{F} = (75.0 \text{ kN})(-\mathbf{i}) = 75.0 \text{ kN}$ opposite to the motion □

or $F = \dfrac{m \Delta v}{\Delta t} = \dfrac{(1500 \text{ kg})(15 \text{ m/s})}{(0.3 \text{ s})} = 75,000 \text{ N}$ □

9.7 **PROBLEM:** A 40-kg child standing on a frozen pond throws a 0.5-kg stone to the east with a speed of 5 m/s. Neglecting friction between child and ice, find the recoil velocity of the child.

SOLUTION: Momentum is conserved for the child-plus-stone system

$p_{1i} + p_{2i} = p_{1f} + p_{2f}$

$0 + 0 = (40 \text{ kg})(v_{1f}) + (0.5 \text{ kg})(5 \text{ m/s})$

$v_{1f} = -\dfrac{(2.5 \text{ kg} \cdot \text{m/s})}{40 \text{ kg}} = 62.5$ mm/s west

9.15   **PROBLEM:**   An estimated force-time curve for a baseball struck

by a bat is shown in Figure 9.15.
From this curve, determine (a) the
impulse delivered to the ball, (b) the
average force exerted on the ball, and
(c) the peak force exerted on the
ball.

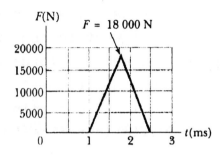

Figure 9.15

SOLUTION:

(a)   Impulse = $\int F dt$ = area under the = F-versus-t graph

$$= \left(\frac{0 + 18000 \text{ N}}{2} \right) (2.5 - 1) \, 10^{-3} \text{ s} = 13.5 \text{ N·s} \quad \square$$

(b) $F_{avg} = \dfrac{\int F dt}{\Delta t} = \dfrac{13.5 \text{ N·s}}{(2.5 - 1) \, 10^{-3} \text{ s}} = 9000 \text{ N} \quad \square$

(c)   From the graph, $F_{max} = 18,000 \text{ N} \quad \square$

9.17   **PROBLEM:**   A 0.15-kg baseball is thrown with a speed of 40 m/s. It is hit straight back at the pitcher with a speed of 50 m/s. (a) What is the impulse delivered to the baseball?   (b) Find the average force exerted by the bat on the ball if the two are in contact for 2×10^{-3} s.   Compare this with the weight of the ball and determine whether or not the impulse approximation is valid in this situation.

SOLUTION: Take the x-axis from pitcher toward batter.   Take initial and final points just before and after bat contacts ball.   Then,

$\mathbf{I} = \Delta\mathbf{p} = \mathbf{p}_f - \mathbf{p}_i$

$\mathbf{I} = (0.15\ \text{kg})(50\ \text{m/s})(-\mathbf{i}) - (0.15\ \text{kg})(40\ \text{m/s})(\mathbf{i})$

$= 13.5\ \text{kg·m/s}\ (-\mathbf{i}) = 13.5\ \text{kg·m/s toward pitcher}\ \square$

$I = \int F dt = F_{avg}\Delta t$

$F_{avg} = I/\Delta t = 13.5\ \text{kg·m/s}(-\mathbf{i})/2 \times 10^{-3}\ \text{s} = 6.75\ \text{kN}(-\mathbf{i})$

The weight of the ball is

$w = mg = (0.15\ \text{kg})(9.8\ \text{m/s}^2) = 1.47\ \text{N down.}$

This is smaller than the force of the bat by 5000 times, so the impulse approximation is good.   The vertical velocity of the ball changes negligibly in 2 ms.

Are you studying for an exam?   Then do this separate problem about the same situation:

Find the work the bat does on the ball.

Solution:

$W = \Delta K = K_f - K_i$

$= \tfrac{1}{2}(0.15\ \text{kg})(50\ \text{m/s})^2 - \tfrac{1}{2}(0.15\ \text{kg})(40\ \text{m/s})^2$

$= 2380\ \text{J}$

Notice how the ball's reversal in direction implies a big momentum change, but does not figure into its energy change.

9.23   **PROBLEM:**   A 10-g bullet is stopped in a 5-kg block of wood. The speed of the bullet-plus-wood combination is immediately after the collision is 0.60 m/s. What was the original speed of the bullet?

SOLUTION: We suppose the block of wood was originally stationary. Total momentum of the bullet + block system is constant over the short time during which relative motion stops:

$$\mathbf{p}_{1i} + \mathbf{p}_{2i} = \mathbf{p}_{1f} + \mathbf{p}_{2f}$$

$$(0.01 \text{ kg})(v_{1i}) + 0 = (0.01 \text{ kg})(0.60 \text{ m/s})\mathbf{i} + (5 \text{ kg})(0.60 \text{ m/s})\mathbf{i}$$

$$\mathbf{v}_{1i} = \frac{(5.01 \text{ kg})(0.60 \text{ m/s})\mathbf{i}}{0.01 \text{ kg}} = (301 \text{ m/s})\mathbf{i} \quad \square$$

9.25 **PROBLEM:** A 1200-kg car, traveling initially with a speed of 25 m/s in an easterly direction, crashes into the rear end of a 9000-kg truck that is moving in the same direction at 20 m/s (Fig 9.25). The velocity of the car immediately after the collision is 18 m/s to the east. (a) What is the velocity of the truck immediately after the collision? (b) How much mechanical energy is lost in the collision? How do you account for this loss in energy?

SOLUTION: Momentum is conserved over the short time of collision:

$p_{1i} + p_{2i} = p_{1f} + p_{2f}$

(1200 kg)(25 m/s) east + (9000 kg)(20 m/s) east

= (1200 kg)(18 m/s) east + (9000 kg) v_{2f}

$$v_{2f} = \frac{(30{,}000 + 180{,}000 - 21{,}600) \text{ kg} \cdot \text{m/s east}}{9000 \text{ kg}}$$

= 20.9 m/s east □

Energy need not be conserved:

$K_{1i} + K_{2i} + \Delta K_{nc} = K_{1f} + K_{2f}$

$\frac{1}{2}$(1200 kg)(25 m/s)^2 + $\frac{1}{2}$(9000 kg)(20 m/s)^2 + ΔK_{nc}

= $\frac{1}{2}$(1200 kg)(18 m/s)^2 + $\frac{1}{2}$(9000 kg)(20.9 m/s)^2

375 kJ + 1.80 MJ + ΔK_{nc} = 194 kJ + 1.97 MJ

ΔK_{nc} = -8.68 kJ

So 8.68 kJ of mechanical energy is lost into thermal energy.

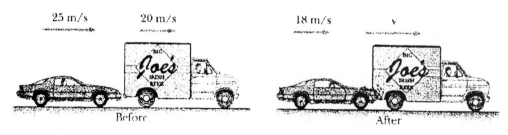

Figure 9.25

9.27 **PROBLEM:** A neutron in a reactor makes an elastic head-on collision with the nucleus of a carbon atom that is initially at rest. (a) What fraction of the neutron's kinetic energy is transferred to the carbon nucleus? (b) If the initial kinetic energy of the neutron is 1 MeV = 1.6×10^{-13} J, find its final kinetic energy and the kinetic energy of the carbon nucleus after the collision. (The mass of the carbon nucleus is about 12 times the mass of the neutron.)

SOLUTION: In this special case of a perfectly elastic head-on collision, we may use the relative velocity equation

$$v_{1i} - v_{2i} = -(v_{1f} - v_{2f})$$

along with conservation of momentum

$$m_1 v_{1i} + m_2 v_{2i} = m_1 v_{1f} + m_2 v_{2f}$$

Let object 1 be the neutron and 2 be the carbon nucleus, with $v_{2i} = 0$ and $m_2 = 12m_1$. Then

$$v_{1i} = -v_{1f} + v_{2f}$$

$$m_1 v_{1i} = m_1 v_{1f} + 12 m_1 v_{2f}$$

Regard v_{1i} as known. The two equations let us solve for the two unknowns v_{1f} and v_{2f}:

$$v_{2f} = v_{1i} + v_{1f}$$

Substituting,

$$v_{1i} = v_{1f} + 12 (v_{1i} + v_{1f})$$

$$-11 v_{1i} = 13 v_{1f} \qquad v_{1f} = -(11/13) v_{1i}$$

$$v_{2f} = v_{1i} - (11/13) v_{1i} = (2/13) v_{1i}$$

The neutron's original kinetic energy is $\frac{1}{2} m_1 v_{1i}^2$. The carbon's final kinetic energy is

$$\frac{1}{2} m_2 v_{2f}^2 = \frac{1}{2} (12 \, m_1) \left(\frac{2}{13} \right)^2 v_{1i}^2 = \left(\frac{48}{169} \right) \left(\frac{1}{2} \right) m_1 v_{1i}^2$$

So $\left(\dfrac{48}{169} \right) = 0.284$ □ of the total energy is transferred.

(b) For the carbon nucleus,

$$K_{2f} = (0.284)(1 \text{ MeV}) = 0.284 \text{ MeV} \ \square$$

The collision is perfectly elastic, so the neutron retains the rest of the energy,

$$K_{1f} = 1 \text{ MeV} - 0.284 \text{ MeV} = 0.716 \text{ MeV} \ \square$$

9.39 **PROBLEM:** A 3-kg mass with an initial velocity of 5**i** m/s
      collides with the sticks to a 2-kg mass with an initial
      velocity of -3**j** m/s. Find the final velocity of the composite
      mass.

      SOLUTION: Momentum is conserved, with both masses having the
      same final velocity:
      $m_1\mathbf{v}_{1i} + m_2\mathbf{v}_{2i} = m_1\mathbf{v}_{1f} + m_2\mathbf{v}_{2f}$

      (3 kg)(5**i** m/s) + (2 kg)(-3**j** m/s) = (3 kg + 2 kg)$\mathbf{v}_f$

      $$\mathbf{v}_f = \frac{15\mathbf{i} - 6\mathbf{j}}{5} \text{ m/s} = (3.00\mathbf{i} - 1.20\mathbf{j}) \text{ m/s}$$

      Make sure you can compute the kinetic energy, which is not a
      vector and not conserved.

      $K_{1i} + K_{2i} = \frac{1}{2}(3 \text{ kg})(5 \text{ m/s})^2 + \frac{1}{2}(2 \text{ kg})(3 \text{ m/s})^2 = 46.5 \text{ J}$

      $K_{1f} + K_{2f} = \frac{1}{2}(5 \text{ kg})(3\mathbf{i} - 1.2\mathbf{j})^2 \text{ m}^2/\text{s}^2$

      $\qquad\qquad = (2.5 \text{ kg})(9 + 1.44) \text{ m}^2/\text{s}^2 = 26.1 \text{ J}$

9.41 **PROBLEM:** A billiard ball moving at 5 m/s strikes a stationary
      ball of the same mass. After the collision, the first ball
      moves at 4.33 m/s, at an angle of 30° with respect to its
      original line of motion. Assuming an elastic collision (and
      ignoring friction and rotational motion), find the magnitude
      and direction of the struck ball's velocity.

      SOLUTION: Call the mass of each ball $m_1 = m_2 = m$.
      Take the x-axis in the direction of the original motion. Call
      f the direction-angle of the second ball, as in Figure 9.11.
      The x-component of momentum is conserved:
      $m(5 \text{ m/s}) = m(4.33 \text{ m/s})\cos 30° + mv_{2fx}$

      $\mathbf{v}_{2fx} = 1.25 \text{ m/s}$

      The y-component of momentum is conserved

      $0 = m(4.33 \text{ m/s})\sin 30° + mv_{2fy}$

      $\mathbf{v}_{2fy} = -2.17 \text{ m/s}$

      Now

      $\mathbf{v}_{2f} = (1.25 \text{ m/s})\mathbf{i} - (2.17 \text{ m/s})\mathbf{j}$

      $\qquad = \sqrt{1.25^2 + 2.17^2} \text{ m/s at Arctan} \dfrac{-2.17}{1.25}$

      $\mathbf{v}_{2f} = 2.50 \text{ m/s at } -60.0° \ \square$

9.45   **PROBLEM:**   An unstable nucleus of mass 17×10^{-27} kg, initially at rest, disintegrates into three particles. One of the particles, of mass 5.0×10^{-27} kg, moves along the y axis with a velocity of 6×10^6 m/s. Another particle, of mass 8.4×10^{-27} kg, moves along the x axis with a velocity of 4×10^6 m/s. Find (a) the velocity of the third particle and (b) the total energy given off in the process.

SOLUTION: With three particles, the total final momentum is $m_1\mathbf{v}_{1f} + m_2\mathbf{v}_{2f} + m_3\mathbf{v}_{3f}$, and it must be zero to equal the original momentum. The mass of the third fragment is

$(17 - 5 - 8.4)(10^{-27}$ kg$) = 3.6 \times 10^{-27}$ kg

$0 = (5 \times 10^{-27}$ kg$)(6 \times 10^6$ m/s$)\mathbf{j}$
$\qquad + (8.4 \times 10^{-27}$ kg$)(4 \times 10^6$ m/s$)\mathbf{i} + (3.6 \times 10^{-27}$ kg$)\ \mathbf{v}_{3f}$

$$\mathbf{v}_{3f} = \frac{(-3\mathbf{j} - 3.36\mathbf{i})(10^{-20}\ \text{kg} \cdot \text{m} \ / \ \text{s})}{3.6 \times 10^{-27}\ \text{kg}} = (-9.33\mathbf{i} - 8.33\mathbf{j})\ \text{Mm/s} \ \square$$

The original kinetic energy is zero. The final kinetic energy is $K = K_{1f} + K_{2f} + K_{3f}$.

$K = \tfrac{1}{2}(5 \times 10^{-27}$ kg$)(6 \times 10^6$ m/s$)^2$
$+ \tfrac{1}{2}(8.4 \times 10^{-27}$ kg$)(4 \times 10^6$ m/s$)^2$
$\qquad + \tfrac{1}{2}(3.6 \times 10^{-27}$ kg$)(9.33^2 + 8.33^2)(10^{12}\ \text{m}^2/\text{s}^2)$
$= (9 \times 10^{-14}$ J$) + (6.72 \times 10^{-14}$ J$) + (28.2 \times 10^{-14}$ J$)$
$K = 4.39 \times 10^{-13}$ J $\square$

9.53   **PROBLEM:** A uniform piece of sheet
steel is shaped as shown in Figure
9.53.   Compute the *x* and *y* coordinates
of the center of mass of the piece.

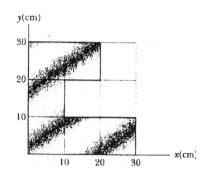

SOLUTION: Think of the sheet as
composed of three sections:

Figure 9.53

1) A rectangle from (0,0) to (30 cm, 10 cm). Its mass is
30 cm x 10 cm x σ, where σ is the mass-per-area of the sheet,
and its center of mass is at (15 cm, 5 cm).
2) A rectangle from (0, 10 cm) to (10 cm, 30 cm) with mass
20 cm x 10 cm x σ and c.m. at (5 cm, 20 cm).
3) A square from (10 cm, 20 cm) to (20 cm, 30 cm), with mass
10 cm x 10 cm x σ and c.m. at (15 cm, 25 cm).
Then the overall c.m. is at

$\mathbf{r}$ = (300σ cm^2)(15**i** + 5**j**) + (200σ cm^3)(5**i** + 20**i**)

$\qquad$ + (100σ cm^3)(15**i** + 25**j**)/[300 + 200 + 100] σ cm^2

$\mathbf{r}$ = (45**i** + 15**j** + 10**i** + 40**j** + 15**i** + 25**j**) cm/6

$\mathbf{r}$ = (70**i** + 80**j**) cm/6 = (11.7 **i** + 13.3 **j**) cm □

If we chose any other division of the original shape, the
answer would be the same.

9.57 **PROBLEM:** Romeo entertains Juliet by playing his guitar at the rear of their boat, which floats in still water. After the serenade, Juliet carefully moves to the rear of the boat (away from shore) to plant a kiss on Romeo's cheek. If the 80-kg boat is facing shore and the 55-kg Juliet moves 2.7 m toward the 77-kg Romeo, how far does the boat nove toward shore?

SOLUTION: No outside forces act on the boat-plus-lovers system, so its momentum is conserved at zero and its center of mass stays fixed: $x_{CMi} = x_{CMf}$. Thus,

$$\frac{M_J x_{JI} + M_R x_{Ri} + M_B x_{Bi}}{M_J + M_R + M_B} = \frac{M_J x_{Jf} + M_R x_{Rf} + M_B x_{Bf}}{M_J + M_R + M_B}$$

$$M_J (x_{Jf} - x_{Ji}) + M_R (x_{Rf} - x_{Ri}) + M_B (x_{Bf} - x_{Bi}) =$$

Romeo sits still in the boat and so undergoes the same displacement $\Delta x_R = \Delta x_B$.

Choose the x-axis pointing away from shore. Then

$$(55 \text{ kg})(\Delta x_J) + (77 \text{ kg} + 80 \text{ kg}) \Delta x_B = 0$$

$$\Delta x_J = -2.85 \Delta x_B$$

As Juliet moves away from shore, the boat and Romeo glide toward shore until the original 2.7 m gap between them is closed:

$$\left| \Delta x_J \right| + \left| \Delta x_B \right| = 2.7 \text{ m}$$

Substituting,

$$2.85 \left| \Delta x_B \right| + \left| \Delta x_B \right| = 2.7 \text{ m}$$

$$\left| D x_B \right| + \frac{2.7 \text{ m}}{3.85} = 0.700 \text{ m}$$

$$\Delta x_B = -0.700 \text{ m} = 0.700 \text{ m toward shore} \quad \square$$

9.59 **PROBLEM:** The first stage of Saturn V space vehicle consumes
fuel at the rate of 1.5 x 10^4 kg/s, with an exhaust speed of
2.6 x 10^3.   (These are approximate figures.)   (a) Calculate the
thrust produced by these engines.   (b) Find the vehicle's
initial acceleration on the launch pad if its initial mass is
3 x 10^6 kg.   [You must include the force of gravity to solve
(b).]

SOLUTION: The force pushing out the exhaust is described by

$$\int F dt = p_f - p_i$$

$$F = \frac{d}{dt} m v_f - 0 = v_f \frac{dm}{dt}$$

The force on the vehicle has the same magnitude but is upward:

$$v_f \frac{dm}{dt} = (2.6 \times 10^3 \text{ m/ s})(1.5 \times 10^4 \text{ kg/s}) = 39.0 \text{ MN} \ \square$$

(b) $\Sigma F = ma$

$$(+39 \times 10^6 \text{ N}) - (3 \times 10^6 \text{ kg})(9.8 \text{ m/s}^2) = (3 \times 10^6 \text{ kg})a$$

$$\frac{(39 \times 10^6 \text{ N}) - (29.4 \times 10^6 \text{ N})}{3 \times 10^6 \text{ kg}} = a = 3.20 \text{ m/s}^2 \ \square$$

9.63 **PROBLEM:** A golf ball (m = 46 g) is struck a blow that makes an angle of 45° with the horizontal. The drive lands 200 m away on a flat fairway. If the golf club and ball are in contact for 7 ms, what is the average force of impact? (Neglect air resistance effects.)

SOLUTION: To find the launch speed of the ball v_0, we treat its projectile motion. Call t its time of flight:

200 m = $v_0 \cos 45° t$

This describes its horizontal component of displacement, and

$0 = v_0 \sin 45° t - \frac{1}{2}(9.8 \text{ m/s}^2) t^2$

describes its vertical motion. When it hits the ground

$$t = \frac{2 v_0 \sin 45°}{9.8 \text{ m/s}^2}$$

Substituting,

$$200 \text{ m} = v_0 \cos 45° \frac{2 v_0 \sin 45°}{9.8 \text{ m/s}^2}$$

$$v_0 = \sqrt{\frac{(200 \text{ m})(9.8 \text{ m/s}^2)}{2 \sin 45° \cos 45°}} = 44.3 \text{ m/s}$$

Now during the collision of club and ball we choose the ball as our system:

$\int \mathbf{F} dt = \mathbf{p}_f - \mathbf{p}_i$

$\mathbf{F}_{avg}(7 \times 10^{-3} \text{ s}) = (46 \times 10^{-3} \text{ kg})(44.3 \text{ m/s})$ at 45° − 0

$\mathbf{F}_{avg}$ = 291 N at 45° □

9.69   **PROBLEM:**   A 70-kg astronaut is working on the engines of her
ship, which is drifting through space with a constant velocity.
Wishing to get a better view of the Universe, she pushes
against the ship and soon finds herself 30 m behind it.
Without a thruster, the only way for the astronaut to return to
the ship is to throw her 0.5-kg wrench directly away from it.
If she throws the wrench with a speed of 20 m/s, how long does
it take her to reach the ship?

SOLUTION: For the astronaut-plus-wrench system, momentum is
conserved at zero:

$\mathbf{p}_{1i} + \mathbf{p}_{2i} = \mathbf{p}_{1f} + \mathbf{p}_{2f}$

$0 = (70 \text{ kg})(\mathbf{v}_{1f}) + (0.5 \text{ kg})(20 \text{ m/s})\mathbf{i}$

$\mathbf{v}_{1f} = (-0.143 \text{ m/s})\mathbf{i}$

She keeps this constant velocity as she traverses the 30 m
distance:

$\mathbf{x} = \mathbf{v}t \quad (-30 \text{ m})\mathbf{i} = (-0.143 \text{ m/s})\mathbf{i}t$

$t = 210 \text{ s} \quad \square$

9.75 **PROBLEM:** A 5-g bullet, moving with an initial speed of 400 m/s, is fired into the passes through a 1-kg block, as in Figure 9.75. The block, initially at rest on a frictionless, horizontal surface, is connected to a spring of force constant 900 N/m. If the block moves 5 cm to the right after impact, find (a) the speed at which the bullet emerges from the block and (b) the mechanical energy lost in the collision.

SOLUTION: While the bullet is passing through the block, momentum is conserved but energy is not, since the spring has no time to inject impulse while internal friction degrades kinetic into thermal energy. While the block squeezes the spring, momentum is not conserved but energy is conserved, since the spring has time to inject impulse to the left, but this force is conservative. Using the conservation laws where they apply and calling the maximum speed of the block v_B, we have momentum conserved in collision:

(0.005 kg)(400 m/s) + 0 = (1 kg)(v_B) + (0.005 kg)v

and energy conservation in compression:

$\frac{1}{2}$(1 kg)(v_B^2) + 0 = 0 + $\frac{1}{2}$(900 N/m)(0.05 m)^2

Then $v_B = \sqrt{\dfrac{1.125 \text{ J}}{0.5 \text{ kg}}} = 1.50$ m/s

and 2 kg·m/s = 1.50 kg m/s + (0.005 kg)v

v = 100 m/s □

For energy in the collision we have

$K_{1i} + K_{2i} + \Delta K_{nc} = K_{1f} + K_{2f}$

$\frac{1}{2}$(0.005 kg)(400 m/s)^2 + 0 + ΔK_{nc}

= $\frac{1}{2}$(0.005 kg)(100 m/s)^2

         + $\frac{1}{2}$(1 kg)(1.5 m/s)^2

400 J + ΔK_{nc} = 25 J + 1.12 J

DK_{nc} = -374 J □ or 374 J lost

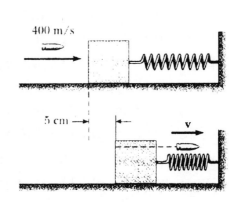

Figure 9.75

CHAPTER 10

QUESTIONS

3   **QUESTION:** Two identically constructed clocks are synchronized. One is put in orbit around the Earth while the other remains on Earth. Which clock runs more slowly? When the moving clock returns to Earth, will the two clocks still be synchronized?

ANSWER: The clock in orbit will run slower. The experiment for clocks carried in airplanes is described on pages 265 - 266. As with the twin "paradox", the extra centripetal acceleration of the orbiting clock makes its history fundamentally different from the clock on Earth.

9   **QUESTION:** List some ways our day-to-day lives would change if the speed of light were only 50 m/s.

ANSWER: Get a "Mr. Tompkins" book by George Gamow for a wonderful fictional exploration of this question. Driving home in a hurry, you push on the gas pedal not to increase your speed by very much, but rather to make the blocks get shorter. Big Doppler shifts in wave frequencies make red lights look green as you approach them and make car horns and car radios useless. High-speed transportation is very expensive, requiring huge fuel purchases. And it is dangerous, as a speeding car can knowck down a building. When you get home hungry for lunch, you have missed dinner. There is a five-day delay in transit when you watch the Olympics in Australia on live TV. It takes ninety-five years for sunlight to reach the Earth. We cannot see the Milky Way; the fireball of the Big Bang surrounds us at the distance of Rigel or Deneb.

13 **QUESTION:** Some distant galaxies, observed as quasars, are receding from us at half the speed of light (or greater). What is the speed of the light we receive from these quasars?

   ANSWER: It is three hundred million meters per second, just the same as the speed of light from a flashlight at rest.

15 **QUESTION:** Imagine an astronaut on a trip to Sirius, which lies 8 lightyears from the Earth. Upon arrival at Sirius, the astronaut finds that the trip lasted 6 years. If the trip was made at a constant speed of 0.8c, how can the 8-lightyear distance be reconciled with the 6-year duration?

   ANSWER: The 8ℓy is the proper length of a rod from here to Sirius, measured by an observer seeing both nearly at rest. The astronaut sees Sirius coming toward her at only 0.8c, but also sees the distance contracted to

   $$8\ell y/\gamma = 8\ell y \sqrt{1 - v^2/c^2} = 8\ell y \sqrt{1 - 0.8^2} = 8\ell y(0.6) = 4.8\ell y$$

   So the travel time she measures on her clocks is, from

   $$v = \frac{d}{t}$$

   $$t = \frac{d}{v} = \frac{4.8 \ \ell y}{0.8c} = 6 \ y$$

CHAPTER 10

PROBLEMS

10.3   **PROBLEM:** In a laboratory frame of reference, an observer notes
that Newton's second law is valid. Show that it is also valid
for an observer moving at a constant speed relative to the
laboratory frame.

SOLUTION: The first observer watches some object accelerate
under applied forces. Call the instantaneous velocity of the
object u_x. The second observer has constant velocity v
relative to the first, which we assume to be small compared
with light, and measures the object to have velocity
$$u_x^1 = u_x - v$$
Then its acceleration is
$$\frac{du_x^1}{dt} = \frac{du_x}{dt} - 0$$
The same as that measured by the first observer. In this non-
relativistic case they measure also the same mass and forces,
so the second observer also confirms that $\Sigma F = ma$.

10.9   **PROBLEM:**   An atomic clock moves at 1000 km/h for one hour as measured by an identical clock on Earth.   How many nanoseconds slow will the moving clock be at the end of the one-hour interval?

SOLUTION: The moving clock measures a proper interval $\Delta t'$ somewhat less than one hour while the observer measures Δt = 1 hour, in

$$\Delta t = \frac{1}{\sqrt{1 - v^2 / c^2}} \Delta t'$$

we want the difference $\Delta t - \Delta t'$.   With $v \ll c$ we can use a binomial expansion

$$\Delta t = (1 - v^2/c^2)^{-1/2} \Delta t'$$

$$\cong \left(1 + \frac{1}{2}\frac{v^2}{c^2} + \text{term order of } \frac{v^4}{c^4}\right)\Delta t'$$

$$\Delta t - \Delta t' \cong \frac{1}{2}\frac{v^2}{c^2} \Delta t' \cong \frac{1}{2}\frac{v^2}{c^2} \Delta t$$

$$= \frac{1}{2}\frac{(10^6 \text{ m}/3600 \text{ s})^2}{(3 \times 10^8 \text{ m/s})^2}(3600 \text{ s}) = 1.54 \text{ ns} \;\square$$

10.13 **PROBLEM:** A muon formed high in the Earth's atmosphere travels at speed $v = 0.99c$ for a distance of 4.6 km before decays into an electron, a neutrino, and an antineutrino ($\mu^- \to e^- + v + \overline{}$). (a) How long does the muon live, as measured in its reference frame? (b) How far does the muon travel, as measured in its frame?

SOLUTION: During its life, the muon sees the Earth lurching up toward it at $0.99c$ for a contracted distance of

(b) $L = \dfrac{L_p}{\gamma} = L_p\left(1 - \dfrac{v^2}{c^2}\right)^{1/2} = (4.6\ \text{km})(1 - 0.99^2)^{\frac{1}{2}}$

$= 0.649\ \text{km} = 649\ \text{m}$ □

So the muon sees the trip as taking time t in

(a) $v = \dfrac{d}{t}$ $t = \dfrac{d}{v} = \dfrac{649\ \text{m}}{0.99(3 \times 10^8\ \text{m/s})} = 2.18\ \text{ms}$ □

By way of contrast, the Earth observer measures dilated lifetime

$\Delta t = \gamma \Delta t' = \dfrac{1}{\sqrt{1 - 0.99^2}}(2.18\ \mu\text{s}) = 15.5\ \mu\text{s}$

or $\dfrac{4.6 \times 10^3\ \text{m}}{0.99(3 \times 10^8\ \text{m/s})} = 15.5\ \mu\text{s}$

10.17 **PROBLEM:** Two jets of material fly away from the center of a radio galaxy in opposite directions. Both jets move at a speed of 0.75c relative to the galaxy. Determine the speed of one jet relative to the other.

SOLUTION: Arbitrarily take the galaxy as the unprimed frame, the jet moving toward the right as the object, and the jet moving toward the left as the primed frame. Then $u_x = 0.75c$ and $v = -0.75c$ so

$$u_x^1 = \frac{u_x - v}{1 - \frac{u_x v}{c^2}} = \frac{0.75c - (-0.75c)}{1 - \frac{0.75c(-0.75c)}{c^2}} = \frac{1.5c}{1 + 0.75^2}$$

$$u_x^1 = 0.960c \quad \square$$

10.25 PROBLEM: An unstable particle at rest breaks into two
fragments of *unequal* mass. The rest mass of the lighter
fragment if 2.5×10^{-28} kg, and that of the heavier fragment is
1.67×10^{-27} kg. If the lighter fragment has a speed of $0.893c$
after the breakup, what is the speed of the heavier fragment?

SOLUTION: Momentum of the pair of fragments is conserved at
its zero original value. With different speeds, the fragments
have different gammas:

$$0 = \gamma_1 m_1 v_1 \mathbf{i} + \gamma_2 m_2 v_2 (-\mathbf{i})$$

$$\frac{1}{\sqrt{1 - v_1^2/c^2}} m_1 v_1 = \frac{1}{\sqrt{1 - v_2^2/c^2}} m_2 v_2$$

$$\frac{1}{\sqrt{1 - 0.893^2}} (2.5 \times 10^{-28} \text{kg})(0.893c) = \frac{v_2}{\sqrt{1 - v_2^2/c^2}}(1.67 \times 10^{-27} \text{ kg})$$

$$0.297c = \frac{v_2}{\sqrt{1 - v_2^2/c^2}}$$

$$0.0882c^2 - 0.0882v_2^2 = v_2^2$$

$$0.0882c^2 = 1.0882v_2^2$$

$$v_2 = \sqrt{0.0882/1.0882}c = 0.285c \quad \square$$

10.29 PROBLEM: A proton moves with the speed of $0.95c$. Calculate
its (a) rest energy, (b) total energy, and (c) kinetic energy.

SOLUTION:

(a) $mc^2 = (1.67 \times 10^{-27} \text{ kg})(3 \times 10^8 \text{ m/s})^2 = 1.50 \times 10^{-10}$ J

$$= (1.50 \times 10^{-10} \text{ J})\left(\frac{1 \text{ eV}}{1.6 \times 10^{-19} \text{ J}}\right) = 939 \text{ MeV} \quad \square$$

(b) $\gamma mc^2 = \left(\frac{1}{\sqrt{1 - 0.95^2}}\right)(939 \text{ MeV}) = 3.20 \times 939 \text{ MeV} = 3.01 \text{ GeV} \quad \square$

(c) $K = (\gamma - 1)mc^2 = 2.20 \times 939 \text{ MeV} = 2.07 \text{ GeV} \quad \square$

10.37 **PROBLEM:** The annual energy requirement of the United States is og the order of 10^{20} J. (a) How many atoms of ^{235}U (with an energy release of 208 MeV per fission event) must be fissioned every second to meet this requirement? (b) How many kilograms of ^{235}U would be required each year?

SOLUTION: The United States converts energy with power

$$10^{20}\frac{J}{y} = 10^{20}\frac{J}{y}\left(\frac{1\ y}{365\ d}\right)\left(\frac{1\ d}{86,400\ s}\right) = 3.17 \times 10^{12}\ W$$

Let n represent the number of fissions each second. Then

$$3.17 \times 10^{12}\frac{J}{s} = n208\frac{MeV}{fission}\left(\frac{10^6\ eV}{1\ MeV}\right)\left(\frac{1.6 \times 10^{-19}\ J}{1\ eV}\right)$$

$n = 9.53 \times 10^{22}$ fission/s $\square$

(b) Each nucleus fissions once, so the rate of material use is

$$9.53 \times 10^{22}\frac{atom}{s}\left(\frac{1\ mole}{6.02 \times 10^{23}\ atoms}\right)\left(\frac{0.235\ kg}{1\ mole}\right)$$

$$= 0.0372\frac{kg}{s} = \left(0.0372\frac{kg}{s}\right)\left(\frac{3.16 \times 10^7\ s}{1\ y}\right)$$

$$= 1.17 \times 10^6\ kg/y\ \square$$

10.40 **PROBLEM:** The oceans have a volume of 317 million cubic miles and contain 1.32×10^{21} kg of water. Of all the hydrogen nuclei in this water, 0.0156% are deuterium. (a) If all of these deuterium nuclei were fused to helium via the first reaction in Equation 10.25, determine the total amount of energy that could be released. (b) Current world energy consumption is about 7×10^{12} W. If consumption were 100 times greater, how many years would the energy supply calculated in (a) last?

SOLUTION: The number of molecules in the ocean is

$$1.32 \times 10^{21} \text{ kg} \left(\frac{1 \text{ mole}}{0.018 \text{ kg}} \right) \left(\frac{6.02 \times 10^{23} \text{ molecule}}{1 \text{ mole}} \right)$$

$$= 4.41 \times 10^{46} \text{ molecules}$$

The number of deuterons is

$$4.41 \times 10^{46} \text{ molecules} \left(\frac{2 \text{ H atoms}}{\text{molecule}} \right) \left(\frac{0.0156 \times 10^{-2} \text{ D}}{\text{H}} \right)$$

$$= 1.38 \times 10^{43} \text{ deuterons}$$

Each fusion event requires two deuterons, so the number of fusions is

$$1.38 \times 10^{43} \text{ D} \left(\frac{1 \text{ fusion}}{2 \text{ D}} \right) = 6.89 \times 10^{42} \text{ fusions}$$

With $^{2}_{1}H + ^{2}_{1}H \rightarrow ^{3}_{2}He + ^{1}_{0}n + 3.27$ MeV, this will put out energy

$$6.89 \times 10^{42} \text{ fusions} \left(\frac{3.27 \text{ MeV}}{\text{fusion}} \right) \left(\frac{10^{6} \text{ eV}}{\text{MeV}} \right) \left(\frac{1.6 \times 10^{-19} \text{ J}}{1 \text{ eV}} \right)$$

Total Energy $= 3.60 \times 10^{30}$ J $\square$

(b) $P = \dfrac{E}{t}$

$$t = \frac{E}{P} = \left(\frac{3.60 \times 10^{30} \text{ J}}{199 \left(7 \times 10^{12} \text{ W} \right)} \right) \left(\frac{W \cdot s}{J} \right) \left(\frac{1 \text{ y}}{3.16 \times 10^{7} \text{ s}} \right) = 163 \text{ My } \square$$

10.43 PROBLEM: The net nuclear reaction inside the Sun is $4p \rightarrow He^4 + DE$. If the rest mass of each proton is 938.2 MeV and the rest mass of the He^4 nucleus is 3727 MeV, calculate the percentage of starting mass that is given off.

SOLUTION: The original rest energy is 4 x 938.2 MeV = 3753 MeV The energy given off is (3753 - 3727) MeV = 25.8 MeV. The fractional energy released is 25.8 MeV/3753 MeV = 0.687%. □

10.45 PROBLEM: The average lifetime of a pi meson in its own frame of reference is 2.6×10^{-8} s. If the meson moves waith a speed of $0.95c$, what are (a) its mean lifetime as measured by an observer on Earth and (b) the average distance it travels before decaying, as measured by an observer on Earth?

SOLUTION: The lifetime 2.6×10^{-8} s is proper time. We measure dilated time

$$\Delta t = \gamma \Delta t' = \frac{1}{\sqrt{1 - 0.95^2}} (2.6 \times 10^{-8} \text{ s}) = 8.33 \times 10^{-8} \text{s} \quad \square$$

(b) $d = vt = 0.95(3 \times 10^8 \text{ m/s})(8.33 \times 10^{-8} \text{s}) = 23.7 \text{ m} \quad \square$

[In its own frame, the pion would see the lab moving only $0.95(3 \times 10^8 \text{ m/s})(2.6 \times 10^{-8}$ s) = 7.41 m, but this is the contracted value of our 23.7 m:

$$L = \frac{L_p}{g} = L_p \sqrt{1 - v^2 / c^2} = (23.7 \text{ m})\sqrt{1 - 0.95^2} = 7.41 \text{ m} \quad \square$$

10.47 PROBLEM: A supertrain (rest length = 100 m) travels at a speed of 0.95c as it passes through a tunnel (rest length = 50 m). As seen by a trackside observer, is the train ever completely within the tunnel? If so, by how much?

SOLUTION: The length of the train as seen by a stationary observer is

$$L = \sqrt{1 - v^2 / c^2}(100 \text{ m})$$

$$L = \sqrt{1 - 0.95^2}(100 \text{ m}) = 31.2 \text{ m}$$

Therefore, the stationary observer "sees" the train entirely within the 50 m tunnel, with 18.8 m to spare. □

10.53 PROBLEM: Imagine that the entire Sun collapses to a sphere of radius R_g such that the work required to remove a small mass m from the surface would be equal to its rest energy mc^2. This radius is called the *gravitational radius* for the Sun. Find R_g. (It is believed that the ultimate fate of many stars is to collapse to their gravitational radii or smaller.)

SOLUTION: The gravitational potential energy of the bit of mass is $-GMm/R_g$, relative to zero energy at infinity, so the work to pull it off is $+GMm/R_g = mc^2$

$$R_g = \frac{GM}{c^2} = \frac{\left(6.67 \times 10^{-11} \text{ N} \cdot \text{m}^2\right)\left(1.99 \times 10^{30} \text{ kg}\right)}{\left(\text{kg}^2\right)\left(3 \times 10^8 \text{ m/s}\right)^2} = 1.47 \text{ km} \quad □$$

CHAPTER 11

QUESTIONS

4   **QUESTION:** What is the magnitude of the angular velocity, ω, of the second hand of a clock? What is the angular acceleration, α, of the second hand?

ANSWER: The second hand of a clock turns this fast:
w = one revolution per minute
w = $(2 \pi$ rad$)/60$ s = 0.105 rad/s $\square$
The motion is clockwise, so the direction of the vector

angular velocity is away from you.

It turns steadily, so ω is constant and $\alpha = 0$ $\square$

5   **QUESTION:** If you see an object rotating, is there necessarily a net torque acting on it?

ANSWER: An object rotates with constant angular momentum when zero total torque acts on it. Consider the Earth.

13   **QUESTION:** A ladder rests inclined
      against a wall, Would you feel safer
      climbing up the ladder if you were told
      that the floor is frictionless but the
      wall is rough, or that the wall is
      frictionless but the floor is rough?
      Justify your answer.

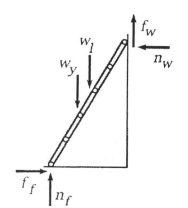

Figure 11.13

      ANSWER: The picture shows the forces on
      the ladder if both wall and floor exert
      friction. If the floor is perfectly
      smooth it can exert no friction force to the right to
      counterbalance the normal force of the wall: the ladder cannot
      stand in equilibrium. On the other hand, a smooth wall can still
      exert a normal force to hold the ladder in equilibrium against
      horizontal motion. The counterclockwise torque of this force
      prevents rotation about the foot of the ladder. So you should
      choose a rough floor.

16   **QUESTION:** As a tetherball winds around a pole, what happens to
      its angular speed? Explain.

      ANSWER: The tetherball's angular velocity increases. If the
      pole is thin, the rope exerts negligible torque on the ball, so
      its angular momentum is constant. As the ball gets closer to the
      pole its moment of inertia decreases, so its angular velocity
      must increase to keep Iw constant. The increased $\frac{1}{2}Iw^2$ energy
      comes from work the string tension does on the ball as the
      tension pulls it inward.

18   **QUESTION:** If the net force acting on a system is zero, then it is necessarily true that the torque on it is also zero?

ANSWER: Forces that add to zero can produce a nonzero total torque. Consider pushing up with one hand on one side of a steering wheel and pulling down equally hard with the other hand on the other side. (A pair of equal-magnitude, oppositely-directed forces applied at different points is called a couple.)

21   **QUESTION:** Vecotr **A** is in the negative *y* direction, and vector **B** is in the negative *x* direction. What are the directions of (a) **A** x **B** and (b) **B** x **A**?

ANSWER: (a) Down-cross-left is away from you
(b) Left-cross-down is toward you

CHAPTER 11

PROBLEMS

11.5   **PROBLEM:**   An electric motor, rotating a workshop grinding wheel at a rate of 100 rev/min, is switched off. Assume constant negative acceleration of magnitude 2 rad/s^2. (a) How long will it take for the grinding wheel to stop? (b) Through how many radians has the wheel turned during the time found in (a)?

SOLUTION: $\omega_0 = 100 \dfrac{\text{rev}}{\text{min}} \left(\dfrac{2\pi \quad \text{rad}}{1 \quad \text{rev}} \right) \left(\dfrac{1 \quad \text{min}}{60 \quad \text{s}} \right) = 10.5$ rad/s

$\omega = 0$ $\alpha = -2$ rad/s^2 $t = ?$ $\theta = ??$

(a) $\omega = \omega_0 + \alpha t$

$$t = \frac{\omega - \omega_0}{\alpha} = \frac{0 - (10.5 \text{ rad/s})}{-2 \text{ rad/s}^2} = +5.24 \text{ s } \square$$

(b) $\omega^2 - \omega_0^2 = 2\alpha(\theta - \theta_0)$

$$\theta - \theta_0 = \frac{\omega^2 - \omega_0^2}{2\alpha} = \frac{0 - (10.5 \text{ rad/s})^2}{2(-2 \text{ rad/s}^2)} = 27.4 \text{ rad } \square$$

Note also in part (b) that since a constant acceleration is acting for time t,

$$\theta = \bar{\omega}t = \left(\frac{10.5 + 0}{2} \right)(5.24) = 27.4 \text{ radians}$$

11.13 **PROBLEM:** A race car travels on a circular track of radius 250 m. If the car moves with a constant speed of 45 m/s, find (a) the angular speed of the car and (b) the magnitude and direction of the car's acceleration.

SOLUTION: (a) $\omega = \dfrac{v}{r} = \dfrac{45 \text{ m/s}}{250 \text{ m}} = 0.180$ rad/s □

(b) With no change in speed, the car has no tangential acceleration. The acceleration acts toward the center

$$a = \frac{v^2}{r} = \frac{(45 \text{ m/s})^2}{250 \text{ m}} = 8.1 \text{ m/s}^2 \ \square$$

11.20 **PROBLEM:** Three particles are connected by rigid rods, of negligible mass, lying along the y axis (Fig. 11.20). The system rotates about the x axis with an angular speed of 2 rad/s. (a) Find the moment of inertia about the x axis and the total kinetic energy evaluated from $\frac{1}{2}I\omega^2$. (b) Find the linear speed of each particle and the total kinetic energy evaluated from $\Sigma\frac{1}{2}m_iv_i^2$.

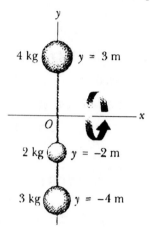

4 kg y = 3 m

2 kg y = -2 m

3 kg y = -4 m

Figure 11.20

SOLUTION:

(a) $I = \Sigma mr^2$

$I = (4 \text{ kg})(3 \text{ m})^2 + (2 \text{ kg})(-2 \text{ m})^2 + (3 \text{ kg})(-4 \text{ m})^2$

$I = 92.0 \text{ kg} \cdot \text{m}^2$

$K = \frac{1}{2}I\omega^2 = \frac{1}{2}(92 \text{ kg} \cdot \text{m}^2)(2 \text{ rad/s})^2 = 184 \text{ J}$ o

(b) The 4-kg mass moves at $v = r\omega = (3 \text{ m})(2 \text{ rad/s}) = 6 \text{ m/s}$

The 2-kg mass has $v = r\omega = (2 \text{ m})(2 \text{ rad/s}) = 4 \text{ m/s}$

and the 3-kg, $v = (4 \text{ m})(2 \text{ rad/s}) = 8 \text{ m/s}$

$K = \Sigma\frac{1}{2}mv^2$

$K = \frac{1}{2}(4 \text{ kg})(6 \text{ m/s})^2 + \frac{1}{2}(2 \text{ kg})(4 \text{ m/s})^2 + \frac{1}{2}(3 \text{ kg})(8 \text{ m/s})^2$

$K = 184 \text{ J}$ ☐

We can call the energy rotational or translational without affecting its amount.

11.21　**PROBLEM:**　Find the net torque on the wheel in Figure 11.21, about the axle through O, if a = 10 cm and b = 25 cm.

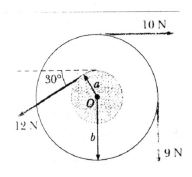

Figure 11.21

SOLUTION:

$\Sigma\tau = \Sigma Fd$

$\Sigma\tau = (+12$ N$)(0.10$ m$) - (10$ N$)(0.25$ m$)$
　　　$- (9$ N$)(0.25$ m$)$

$\Sigma\tau = -3.55$ N·m $\square$ = 3.55 N·m away from you.

Note that the 30° angle is not required for the solution. Note that the 10-N and 9-N forces both produce clockwise torques.

11.25 **PROBLEM:** Two vecotrs are given by $\mathbf{A}$ = -3$\mathbf{i}$ + 4$\mathbf{j}$ and
$\mathbf{B}$ = 2$\mathbf{i}$ + 3$\mathbf{j}$. Find (a) $\mathbf{A} \times \mathbf{B}$ and (b) the angle between $\mathbf{A}$ and $\mathbf{B}$

SOLUTION:

(a) $\mathbf{A} \times \mathbf{B}$ = (-3$\mathbf{i}$ + 4$\mathbf{j}$) x (2$\mathbf{i}$ + 3$\mathbf{j}$)

= (-6$\mathbf{i}$ x $\mathbf{i}$) - (9$\mathbf{i}$ x $\mathbf{j}$) + (8$\mathbf{j}$ x $\mathbf{i}$) + (12$\mathbf{j}$ x $\mathbf{j}$)

= 0 - 9$\mathbf{k}$ + 8(-$\mathbf{k}$) + 0 = -17.0$\mathbf{k}$ □

(b) Since $|\mathbf{A} \times \mathbf{B}| = |\mathbf{A}||\mathbf{B}||\sin\theta|$,

$$\theta = \text{Arcsin}\frac{|\mathbf{A} \times \mathbf{B}|}{AB} = \text{Arcsin}\frac{17}{\sqrt{3^2 + 4^2}\sqrt{2^2 + 3^2}}$$

$$\theta = \text{Arcsin}\frac{17}{18} = 70.6° \quad □$$

To solidify your learning, review taking the dot product
of these same two vectors and finding the angle between
them from it:

$\mathbf{A} \cdot \mathbf{B}$ = (-3$\mathbf{i}$ + 4$\mathbf{j}$) $\cdot$ (2$\mathbf{i}$ + 3$\mathbf{j}$)= -6 + 12 = +6.00

$\mathbf{A} \cdot \mathbf{B}$ = $|\mathbf{A}||\mathbf{B}|$ cosq

$$\theta = \text{Arccos}\frac{|\mathbf{A} \times \mathbf{B}|}{AB} = \text{Arccos}\frac{6}{\sqrt{3^2 + 4^2}\sqrt{2^2 + 3^2}} = 70.6°$$

If the angle between vectors is greater than 90°, then
$\mathbf{A} \cdot \mathbf{B}$ will be negative and Arccos ($\mathbf{A} \cdot \mathbf{B}$/AB) will tell you
the angle correctly, but Arcsin($|\mathbf{A} \cdot \mathbf{B}|$/AB) will not.

11.31   **PROBLEM:**   A model airplane whose mass is 0.75 kg is tethered by a wire so that it flies in a circle 30 m in radius.   The airplane engine provides a thrust of 0.80 N perpendicular to the tethering wire.   (a) Find the torque that the engine thrust produces about the center of the circle.   (b) Find the angular acceleration of the airplane when it is in level flight.   (c) Find the linear acceleration of the airplane tangent to its flight path.   Neglect air drag.

SOLUTION: (a) $\tau = Fd = (0.8 \text{ N})(30 \text{ m}) = 24.0 \text{ N·m}$ □

(b) $I = mr^2 = (0.75 \text{ kg})(30 \text{ m})^2 = 675 \text{ kg·m}^2$

$\Sigma\tau = I\alpha$

$\alpha = \dfrac{\Sigma\tau}{I} = \dfrac{24 \text{ N·m}}{675 \text{ kg·m}^2} = 0.0356 \text{ rad/s}^2$ □

(c) $a = r\alpha = (30 \text{ m})(0.0356/\text{s}^2) = 1.07 \text{ m/s}^2$ □

We could also find this linear acceleration from

$\Sigma F = ma$

$a = \dfrac{\Sigma F}{m} = \dfrac{0.8 \text{ N}}{0.75 \text{ kg}} = 1.07 \text{ m/s}^2$

11.33   **PROBLEM:**   The position vector of a 2-kg particle is given as a function of time by $\mathbf{r} = (6\mathbf{i} + 5t\mathbf{j})$ m.   Determine the angular momentum of the particle as a function of time.

SOLUTION: The velocity of the particle is

$$\mathbf{v} = \frac{d\mathbf{r}}{dt} = \frac{d}{dt}(6\mathbf{i} + 5t\mathbf{j})\text{m} = 5\mathbf{j} \text{ m/s}$$

The angular momentum is

$$\mathbf{L} = \mathbf{r} \times \mathbf{p} = m\mathbf{r} \times \mathbf{v} = (2 \text{ kg})(6\mathbf{i} + 5t\mathbf{j})\text{m} \times 5\mathbf{j} \text{ m/s}$$

$$= (60 \text{ kg·m}^2/\text{s})\mathbf{i} \times \mathbf{j} + (50t \text{ kg·m}^2/\text{s})\mathbf{j} \times \mathbf{j}$$

$$= 60\mathbf{k} \text{ kg·m}^2/\text{s} + 0 = 60.0\mathbf{k} \text{ kg·m}^2/\text{s} \ \square$$

It is constant in time.

11.35   **PROBLEM:**   A light, rigid rod 1 m in length rotates in the *xy* plane about a pivot through the rod's center.   Two particles with masses of 4 kg and 3 kg are connected to its ends (Fig. 11.35).   Determine the angular momentum of the system about the origin at the instant the speed of each particle is 5 m/s.

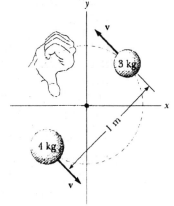

Figure 11.35

   SOLUTION: The moment of inertia of the compound object about its center is

$$I = \Sigma mr^2 = (4 \text{ kg})(0.5 \text{ m})^2 + (3 \text{ kg})(0.5 \text{ m})^2$$

$$I = 1.75 \text{ kg·m}^2$$

Its angular speed is

$$\omega = v/r = (5 \text{ m/s})/(0.5 \text{ m}) = 10 \text{ rad/s toward you.} \quad \text{So its}$$

angular momentum is

$$\mathbf{L} = I\omega = (1.75 \text{ kg·m}^2)(10 \text{ rad/s}) \text{ toward you}$$

$$\mathbf{L} = 17.5 \text{ kg·m}^2/\text{s toward you} \ \square$$

We could also compute this from

$$\mathbf{L} = \Sigma m\mathbf{r} \times \mathbf{v}$$

$$\mathbf{L} = (4 \text{ kg})(0.5 \text{ m})(5 \text{ m/s}) \text{ toward you}$$
$$\qquad + (3 \text{ kg})(0.5 \text{ m})(5 \text{ m/s}) \text{ toward you}$$

$$\mathbf{L} = 17.5 \text{ kg·m}^2/\text{s toward you}$$

11.41   **PROBLEM:**   A woman whose mass is 60 kg stands at the rim of a horizontal turntable having a moment of inertia of 500 kg·m^2 and a radius of 2 m. The system is initially at rest, and the turntable is free to rotate about a frictionless, vertical axle through its center. The woman then starts walking around the rim in a clockwise direction (looking downward), at a constant speed of 1.5 m/s relative to the Earth. (a) In what direction and with what angular speed does the turntable rotate? (b) How much work does the woman do to set the system into motion?

SOLUTION: (a)   The table turns opposite to the way the woman walks, so its angular momentum cancels that of the woman. From conservation of angular momentum we have

$$L_f = L_i = 0$$

so, $L_f = I_w \omega_w + I_{table}\omega_{table} = 0$

and $\omega_{table} = -\dfrac{I_{table}}{I_{table}} \omega_{woman} = -\dfrac{m_w r^2}{I_{woman}} \times \dfrac{v_{woman}}{r}$

$$\omega_{table} = -\dfrac{(60 \text{ kg})(2 \text{ m})^2 (1.5 \text{ m/s})}{500 \text{ kg} \cdot \text{m}^2} = -0.36 \text{ rad/s}$$

or $\omega_{table} = 0.36$ rad/s (counterclockwise) □

(b)   Work done = ΔK

$$W = K_f - 0 = \frac{1}{2} m_{woman} v_{woman}^2 + \frac{1}{2} I \,\omega_{table}^2$$

$$W = \frac{1}{2}(60\,\text{kg})(1.5\,\text{m/s})^2 + \frac{1}{2}\left(500\,\text{kg}\cdot\text{m}^2\right)(0.36\,\text{rad/s})^2$$

$$W = 99.9 \text{ J} \; □$$

You may have to do a lot of explaining on your next exam: Why is angular momentum conserved? Because the axle exerts no torque on the woman-plus-turntable system; only torques from outside can change the total angular momentum. Whay is mechanical energy not conserved? The internal forces of the woman pushing backward on the turntable, and the turntable pushing forward on her, do positive work on both, converting chemical into kinetic energy. Is linear momentum conserved? No. If the woman starts walking north, she pushes south on the turntable. Its axle holds it still against linear motion by pushing north on it, and this outside force delivers northward linear momentum into the system.

11.47 **PROBLEM:** The top in Figure 11.47 has a moment of inertia of 4.0×10^{-4} kg $\cdot$ m^2 and is initially at rest. It is free to rotate about the stationary axis AA'. A string, srapped around a peg along the axis of the top, is pulled in such a manner as to maintain a constant tension of 5.57 N. If the string does not slip while it is unwound from the peg, what is the angular speed of the top after 80 cm of string has been pulled off of the peg? (*Hint:* Consider the work done.)

SOLUTION: The string does work on the top to increase its kinetic energy:

$$(K_{\text{trans}} + K_{\text{rot}} + U_g)_i + DK_{nc} = (K_{\text{trans}} + K_{\text{rot}} + U_g)_f$$

$$Fs \cos q = \tfrac{1}{2} I \omega^2$$

$$(5.57 \text{ N})(0.80 \text{ m}) \cos 0° = \tfrac{1}{2}(4 \times 10^{-4} \text{kg} \cdot \text{m}^2)\omega^2$$

$$\omega = \sqrt{\frac{2 \times 4.46 \text{ J}}{4 \times 10^{-4} \text{ kg} \cdot \text{m}^2}} = 149 \text{ rad/s} \ \square$$

We have followed the hint and solved the problem most efficiently. We could also solve it from Newton's second law for rotation, in several steps. Call r the radius of the drum on which the string is wound. Then the torque of the string is

$\mathbf{t} = \mathbf{r} \times \mathbf{F} = r(5.57 \text{ N}) \sin 90°$ down. From $\Sigma \tau = I\alpha$, its angular acceleration is

$$\alpha = \frac{\Sigma \tau}{I} = \frac{r(5.57 \text{ N}) \text{ down}}{4 \times 10^{-4} \text{ kg} \cdot \text{m}^2} = 13{,}900r \ \frac{1}{\text{m} \cdot \text{s}^2} \text{ down}$$

It turns through angle

$\theta = s/r = (0.8 \text{ m})/r$,

according to

$$\omega^2 - \omega_0^2 = 2\alpha\theta$$

$$\omega^2 - 0 = 2\left(13{,}900 \, r \ \frac{1}{\text{m} \cdot \text{s}^2}\right)\left(\frac{0.80 \text{ m}}{r}\right)$$

$\omega = 149$ rad/s

11.52 **PROBLEM:**   (a) A uniform solid disk of
radius R and mass M is free to rotate on
a frictionless pivot through a point on
its rim (Fig.11.52).   If the disk is
released from rest in the position shown
by the green circle, what is the speed
of its center of mass when the disk
reaches the position indicated by the
dashed circle?   (b) What is the speed of
the lowest point on the disk in the dashed position?   (c)
Repeat part (a) using a uniform hoop instead of a disk.

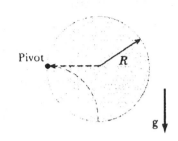

Figure 11.52

SOLUTION: The pivot being frictionless, energy is conserved
during the motion.   To identify the original gravitational
energy, think of the height through which the center of mass
falls.   To identify the final kinetic energy, you can think of
the motion as rolling on a one-point track, with center-of-
mass translational kinetic energy plus energy of rotation
about the center of mass:

$$(K_{trans} + K_{rot} + U_g)_i + \Delta K_{nc} = (K_{trans} + K_{rot} + U_g)_f$$

$$0 + 0 + MgR + 0 = \tfrac{1}{2} M v_{cm}^2 + \tfrac{1}{2} I \,\omega^2 + 0$$

we have $\omega = v_{cm}/R$ and $I = \tfrac{1}{2}MR^2$ for a disk, so

$$MgR = \tfrac{1}{2} M v_{cm}^2 + \tfrac{1}{2}\left(\tfrac{1}{2} MR^2\right) v_{cm}^2 / R^2 = \tfrac{3}{4} M v_{cm}^2$$

$$v_{cm} = \sqrt{4gR / 3} \quad \square$$

(b)   Twice as far from the axis, the bottom of the disk moves twice
as fast through space:

$$\omega = v/r = v_{cm}/R = \sqrt{4gR / 3} \, / \, R$$

$$v = \omega r$$

$$v_b = \omega 2R = \left(\sqrt{4gR / 3} \, / \, R\right) 2R = 2\sqrt{4gR / 3} \quad \square$$

11.52 (cont.)

(c) The hoop has larger moment of inertia, so it turns slower. From the conservation of energy equation above

$$MgR = \tfrac{1}{2} M v_{cm}^2 + \tfrac{1}{2}\left(MR^2\right) v_{cm}^2 / R^2 = M v_{cm}^2$$

$$v_{cm} = \sqrt{gR} \quad \square$$

We could not solve this problem by using $\Sigma\tau = I\alpha$ to find the angular acceleration and then

$\omega^2 - \omega_0^2 = 2\alpha\theta$: with $\theta = \pi/2$ to find ω. This method would give a wrong answer because a is not constant. The lever arm of the weight changes.

11.53 PROBLEM: A grinding wheel is in the form of a uniform solid disk of radius 7 cm and mass 2 kg. It starts from rest and accelerates uniformly under the action of the constant torque of 0.6 N · m that the motor exerts on the wheel. (a) How long does the wheel take to reach its final operating speed of 1200 rpm? (b) Through how many revolutions does the grindstone turn while accelerating?

SOLUTION: The moment of inertia of the wheel is

$$I = \tfrac{1}{2}Mr^2 = \tfrac{1}{2}(2 \text{ kg})(0.07 \text{ m})^2 = 0.0049 \text{ kg} \cdot \text{m}^2$$

From $\Sigma\tau = I\alpha$, its angular acceleration is

$$a = \frac{\Sigma\tau}{I} = \frac{0.6 \text{ N} \cdot \text{m}}{0.0049 \text{ kg} \cdot \text{m}^2} = 122 \text{ rad/s}^2$$

Now, $\omega = \omega_0 + \alpha t$, so

$$t = \frac{\omega - \omega_0}{\alpha} = \left(1200 \frac{\text{rev}}{\text{min}}\right)\left(\frac{2\pi \text{ rad}}{\text{rev}}\right)\left(\frac{\text{min}}{60 \text{ s}}\right)\left(\frac{\text{s}^2}{122 \text{ rad}}\right) = 1.03 \text{ s} \; \square$$

(b) $\omega^2 - \omega_0^2 = 2\alpha(\theta - \theta_0)$

$$\theta - 0 = \frac{\omega^2 - \omega_0^2}{2\alpha} = \frac{(126 \text{ rad/s})^2 - 0}{2 \times 122 \text{ rad/s}^2} = 64.5 \text{ rad}$$

$$\theta = (64.5 \text{ rad})\left(\frac{1 \text{ rev}}{2\pi \text{ rad}}\right) = 10.3 \text{ rev} \; \square$$

We could alternatively use the work-energy theorem as expressed in equation 11.35. With t constant,

$$\int \tau d\theta = \tau \int d = \tau(\theta - \theta_0) = \tfrac{1}{2}I\omega^2 - \tfrac{1}{2}I\omega_0^2$$

$$(0.6 \text{ N} \cdot \text{m})\theta - 0 = \tfrac{1}{2}(0.0049 \text{ kg} \cdot \text{m}^2)(126 \text{ rad/s})^2 - 0$$

$$\theta = \frac{38.77 \text{ J}}{0.6 \text{ N} \cdot \text{m}} = 64.5 \text{ rad}$$

11.59 **PROBLEM:** A small, solid sphere of mass m and radius r rolls, without slipping, along the track shown in Figure 11.59. If it starts from rest at the top of the track at a height h, (a) what is the minimum value of h (in terms of the radius of the loop R) such that the sphere completes the loop? You may assume that h and R are much larger than r. (b) What are the force components on the sphere at point P if $h = 3R$?

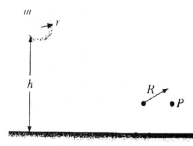

Figure 11.59

SOLUTION: (a) We will not make the simplifying assumption $r \ll R$. We will assume the slope of the track is small, or that h is measured to the lowest point of the ball, so that the original height of its center of mass is $h + r$. If h has its minimum value, the ball will nearly leave the track at the top of the loop: the normal force on it will here be zero. So analysis of the forces on it here reads $\Sigma F = ma$

$$-mg = -m\frac{v^2}{(R - r)},$$ where $(R - r)$ is the radius of the circle in which its center of mass moves. So the speed at the top is

$v = \sqrt{g(R - r)}$ and the angular speed is $w = v/r = \sqrt{g(R - r)}\,/r$. Next, consider conservation of energy between the starting point and the top of the loop:

$$(K_{trans} + K_{rot} + U_g)_i + \Delta K_{nc} = (K_{trans} + K_{rot} + U_g)_f$$

$$0 + 0 + mg(h + r) + 0 = \tfrac{1}{2}mv^2 + \tfrac{1}{2}I\omega^2 + mg(2R - r)$$

$$mgh + mgr = \frac{1}{2}mg(R - r) + \frac{1}{2}\left(\frac{2}{5}\right)\frac{(mr^2 g)(R - r)}{r^2} + mg2R - mgr$$

$$h = -r + \frac{1}{2}R - \frac{1}{2}r + \frac{1}{5}R - \frac{1}{5}r + 2R - r = 2.70(R - r)$$

11.59 **(cont.)**

(b) We first find the speed at point P by using conservation of energy:

$$(K_{trans} + K_{rot} + U_g)_i + \Delta K_{nc} = (K_{trans} + K_{rot} + U_g)_f$$

$$0 + 0 + mg(3R + r) + 0 = \tfrac{1}{2}mv^2 + \tfrac{1}{2}Iw^2 + mgR$$

$$mg3R + mgr = \frac{1}{2}mv^2 + \frac{1}{2}\left(\frac{2}{5}\right)\left(mr^2\right)\left(\frac{v}{r}\right)^2 + mgR$$

$$mg(2R + r) = \left(\frac{1}{2} + \frac{1}{5}\right)\left(mv^2\right) = \frac{7}{10}mv^2$$

Now consider the forces on the sphere at P. The normal force horizontally to the left provides the centripetal force:

$$\Sigma F_x = ma_x \qquad -n = -\frac{mv^2}{(R - r)} = -\left(\frac{10}{7}\right)\left(\frac{mg(2R + r)}{R - r}\right)$$

While gravity acts unopposed to cause a tangential acceleration

$$\Sigma F_y = ma_y \qquad -mg = -ma_y$$

So the force on the ball at P is

$$\mathbf{F}_P = -\left(\frac{10}{7}\right)\left(\frac{mg(2R + r)}{R - r}\right)\mathbf{i} - mg\mathbf{j} \quad \square$$

11.63   **PROBLEM:**   Two blocks, as shown in
Figure 11.63, are connected by a
string of negligible mass passing
over a pulley of radius 0.25 m and
moment of inertia I.   The block on
the incline is moving up with a
constant acceleration of 2 m/s^2.
(a) Determine T_1 and T_2, the
tensions in the two parts of the string.   (b) Find the moment
of inertia of the pulley.

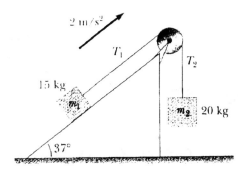

Figure 11.63

SOLUTION:
The 15-kg block weighs $w = mg = $ (15 kg)(9.8 m/s^2) = 147 N.
We suppose the incline is frictionless.   Taking the x-axis up
the incline, we have

$\Sigma F_x = ma_x$ (-147 N)sin 37° + T_1 = (15 kg)(2 m/s^2)

T_1 = 118 N □

For the 20-kg block we have $\Sigma F_y = ma_y$, or

+T_2 - (20 kg)(9.8 m/s^2) = (20 kg)(-2 m/s^2).   So T_2 = 156 N □
Now for the pulley

$\alpha = \dfrac{a}{r} = \dfrac{-2 \ \text{m/s}^2}{0.25 \ \text{m}} = -8$ rad/s^2, the negative sign indicating

clockwise

$\Sigma \tau = I\alpha$, or (+118 N)(0.25 m) - (156 N)(0.25 m) = I(-8 rad/s^2)

$I = \dfrac{9.38 \ \text{N} \cdot \text{m}}{8 \ \text{rad/s}^2} = 1.17$ kg $\cdot$ m^2 □

CHAPTER 12

QUESTIONS

1   **QUESTION:** If the gravitational force on an object is directly
    proportional to its mass, why don't large masses fall with greater
    acceleration than small ones?

    ANSWER: Suppose one object has five times the mass of another. It
    feels five times the force of gravity. As described by Newton's
    second law, a five-times-larger force is required to give the
    object the same acceleration.

6   **QUESTION:**   Explain why it takes more fuel for a spacecraft to travel from the Earth to the Moon than for the return trip. Estimate the difference.

ANSWER: To travel between Earth and Moon, a distance d, a motor must boost the spacecraft over the point of zero total gravitational field in between.   Call x the distance of this point from Earth.   To cancel, Earth and Moon must here produce equal-size fields:

$$\frac{GM_E}{x^2} = \frac{GM_m}{(d-x)^2} \qquad\qquad \frac{M_E}{M_m}(d-x)^2 = x^2$$

$$\sqrt{\frac{5.98 \times 10^{24} \text{ kg}}{7.36 \times 10^{22} \text{ kg}}}(d-x) = x$$

$$9.01d - 9.01x = x$$

$$x = \frac{9.01}{10.01}d = 0.900(3.84 \times 10^8 \text{ m}) = 3.46 \times 10^8 \text{ m}$$

The difference in energy between this point and the Earth's surface, for each kilogram of craft, is approximately,

$$\frac{-GM_E (1 \text{ kg})}{x} - \left(\frac{-GM_E (1 \text{ kg})}{R_e}\right)$$

$$\Delta E_1 = \left(\frac{\left(6.67 \times 10^{-11} \text{ N} \cdot \text{m}^2\right)\left(5.98 \times 10^{24} \text{ kg}\right)(1 \text{ kg})}{\text{kg}^2}\right)$$

$$\left(\frac{1}{6.37 \times 10^6 \text{ m}} - \frac{1}{3.46 \times 10^8 \text{ m}}\right)$$

$$\Delta E_1 = 6.15 \times 10^7 \text{ J}$$

Similarly, the difference in energy between this point and the Moon's surface, for each kilogram of craft, is approximately,

$$\left(\frac{\left(6.67 \times 10^{-11} \text{ N} \cdot \text{m}^2\right)\left(7.36 \times 10^{22} \text{ kg}\right)(1 \text{ kg})}{\text{kg}^2}\right)\left(\frac{1}{1.74 \times 10^6 \text{ m}} - \frac{1}{3.84 \times 10^7 \text{ m}}\right)$$

$$\Delta E_2 = 2.69 \times 10^6 \text{ J, smaller by}$$

$$\frac{\Delta E_1}{\Delta E_2} = \frac{6.15 \times 10^7 \text{ J}}{2.69 \times 10^6 \text{ J}} = 22.8 \text{ times}$$

14   **QUESTION:** Henry Cavendish, in his 1798 experiment, was said to
      have "weighed the Earth." Explain this statement.

      ANSWER: The Earth creates a gravitational field at its surface
      according to $g = GM_e / R_e^2$. The factors g and R_e were known, so as
      soon as Cavendish measured G, he could compute the mass of the
      Earth.

20   **QUESTION:** Discuss the similarities and differences between the
      classical description of planetary motion and the Bohr model of the
      hydrogen atom.

      ANSWER: Both a planet and the electron in the Bohr model of the
      hydrogen atom move in circles around a much more massive attracting
      center. The planet can have any amount of energy or of angular
      momentum, but the electron can have only one of a set of certain
      special quantized energies, only an integer multiple of h for its
      angular momentum. The planet is stable in any orbit, but the
      electron will lose energy spontaneously in discrete steps until it
      is in its lowest state.

CHAPTER 12

PROBLEMS

12.3 **PROBLEM:** The gravitational field on the surface of the Moon is about one-sixth that on the surface of the Earth. If the radius of the Moon is about one-quarter that of the Earth, find the ratio of the average mass density of the Moon to the average mass density of the Earth.

SOLUTION: The gravitational field at the surface of Earth or Moon is given by $g = GM/R^2$. With $\rho = M/V$, $M = \rho V = \rho \frac{4}{3} \pi R^3$, this is $g = G\rho \frac{4}{3} \pi R^3 / R^2 = G\rho \frac{4}{3} \pi R$. Dividing this equation for the Moon by its counterpart for the Earth gives

$$\frac{g_{Moon}}{g_{Earth}} = \frac{G \, \rho_{Moon} \, 4\pi \, R_{Moon} / 3}{G \, \rho_{Earth} \, 4\pi \, R_{Earth} / 3}$$

$$\frac{1}{6} = \left(\frac{\rho_{Moon}}{\rho_{Earth}}\right)\left(\frac{1}{4}\right)$$

$$\frac{\rho_{Moon}}{\rho_{Earth}} = \frac{4}{6} = \frac{2}{3} \quad \square$$

12.7   **PROBLEM:**   The *Explorer VIII* satellite, placed into orbit Nov. 3, 1960, to investigate the ionosphere, had the following orbit parameters: perihelion altitude, 459 km; aphelion altitude, 2289 km (both distances above the Earth's surface); period, 112.7 min. Find the ratio v_p/v_a.

SOLUTION: The satellite moves with constant angular momentum. Equating it at aphelion and perihelion gives

$$mr_a v_a \sin 90° = mr_p v_p \sin 90°$$

$$\frac{v_p}{v_a} = \frac{r_a}{r_p} = \frac{6370 \text{ km} + 2289 \text{ km}}{6370 \text{ km} + 459 \text{ km}} = 1.27 \ \square$$

12.11   **PROBLEM:**   A "synchronous" satellite, which always remains above the same point on a planet's equator, is put in orbit around Jupiter to study the famous red spot. Jupiter rotates once every 9.9 h.   Use the data of Table 12.2 to find the altitude of such an orbiting satellite on Jupiter.

SOLUTION: The gravitational force is the centripetal force:

$$\frac{GM_s M_J}{r^2} = \frac{M_s v^2}{r} = \left(\frac{M_s}{r}\right)\left(\frac{2\pi r}{T}\right)^2$$

$$GM_J T^2 = 4\pi^2 r^3$$

$$r = \left(\frac{Gm_J T^2}{4 p^2}\right)^{1/3} = \left(\frac{(6.67 \times 10^{-11}\,\text{N} \cdot \text{m}^2)(1.90 \times 10^{27}\,\text{kg})(9.9 \times 3600\,\text{s})^2}{(\text{kg}^2)(4\pi^2)}\right)^{1/3}$$

$$r = 15.98 \times 10^7 \text{ m}$$

$$\text{altitude} = (15.98 \times 10^7 \text{ m}) - (6.99 \times 10^7 \text{ m}) = 8.99 \times 10^7 \text{ m} \ \square$$

12.17 **PROBLEM:** A spaceship is fired from Earth's surface with an initial speed of 2.0×10^4 m/s. What will its speed be when it is very far from the Earth? (Neglect friction.)

SOLUTION: Energy is conserved between surface and the distant point:

$$(K + U_g)_i + W_{nc} = (K + U_g)_f$$

$$\frac{1}{2}mv_i^2 - \frac{GM_e m}{R_e} + 0 = \frac{1}{2}mv_f^2 - \frac{GM_e m}{\infty}$$

$$v_f^2 = v_i^2 - \frac{2GM_e}{R_e}$$

$$v_f^2 = (2 \times 10^4 \text{ m/s})^2 - \frac{2(6.67 \times 10^{-11} \text{ N} \cdot \text{m}^2)(5.98 \times 10^{24} \text{ kg})}{\text{kg}^2(6.37 \times 10^6 \text{ m})}$$

$$v_f^2 = 4 \times 10^8 \text{ m}^2/\text{s}^2 - 1.25 \times 10^8 \text{ m}^2/\text{s}^2 = 2.75 \times 10^8 \text{ m}^2/\text{s}^2$$

$$v_f = 1.66 \times 10^4 \text{ m/s } \square$$

12.21 **PROBLEM:** A satellite moves in a circular orbit just above the surface of a planet. Show that the orbital speed, v, and the escape speed v_{esc} of the satellite are related by the expression $v_{esc} = \sqrt{2}\,v$.

SOLUTION: Call M the mass of the planet and R its radius. For the orbiting "treetop satellite", $SF = ma$ reads

$$\frac{GMm}{R^2} = \frac{mv^2}{R}$$

$$v = \sqrt{\frac{GM}{R}}$$

For conservation of energy for an object launched with escape velocity,

$$\frac{1}{2}mv_{esc}^2 - \frac{GMm}{R} = 0$$

$$v_{esc} = \sqrt{\frac{2GM}{R}}$$

Thus, $v_{esc} = \sqrt{2}\,v$ □

12.31   **PROBLEM:**   A hydrogen atom is in its first excited state ($n = 2$). Using the Bohr theory of the atom, calculate (a) the radius of the orbit, (b) the linear momentum of the electron, (c) the angular momentum of the electron, (d) the kinetic energy, (e) the potential energy, and (f) the total energy.

SOLUTION:

(a)   From Equations 12.20 and 12.21,

$$r_n = n^2 a_0$$

$$r_2 = 4 \times 0.0529 \text{ nm} = 0.212 \text{ nm} \quad \square$$

We can more easily do (c) before (b):

$$mvr = n\hbar = \frac{2(6.63 \times 10^{-34} \text{ J} \cdot \text{s})}{2\pi} = 2.11 \times 10^{-34} \text{ J} \cdot \text{s} \quad \square$$

$$mv = \frac{mvr}{r} = \frac{2.11 \times 10^{-23} \text{ J} \cdot \text{s}}{0.212 \times 10^{-9} \text{ m}} = 9.95 \times 10^{-25} \text{ kg} \cdot \text{m/s}$$

(d) $$K = \frac{1}{2}mv^2 = \frac{m^2v^2}{2m} = \frac{p^2}{2m} = \frac{(9.95 \times 10^{-25} \text{ kg} \cdot \text{m/s})^2}{2(9.11 \times 10^{-31} \text{ kg})}$$

$$K = 5.44 \times 10^{-19} \text{ J} = (5.44 \times 10^{-19} \text{ J})\left(\frac{1 \text{ eV}}{1.6 \times 10^{-19} \text{ J}}\right)$$

$$K = 3.40 \text{ eV} \quad \square$$

(e) $$U = \frac{-k_e e^2}{r} = \frac{(-8.99 \times 10^9 \text{ N} \cdot \text{m}^2)(1.6 \times 10^{-19} \text{ C})^2}{c^2(0.212 \times 10^{-9} \text{ m})} = -6.80 \text{ eV} \quad \square$$

(f) $$E = K + U = 3.40 \text{ eV} - 6.80 \text{ eV} = -3.40 \text{ eV} \quad \square$$

$$\text{or } E = (-13.6 \text{ eV})/n^2 = (-13.6 \text{ eV})/2^2 = -3.40 \text{ eV}$$

12.35 **PROBLEM:**   A cylindrical habitat in space, 6 km in diameter and 30 km long, has been proposed (by G. K. O'Neill, 1974). Such a habitat would have cities, land, and lakes on the inside surface and air and clouds in the center. This would all be held in place by rotation of the cylinder about its long axis. How fast would the cylinder have to rotate to imitate a 1-g gravity field at the walls of the cylinder?

SOLUTION: For a 6-km diameter cylinder, $r = 3000$ m and to simulate $1g = 9.8$ m/s^2, we have

$$g = \frac{v^2}{r} = \omega^2 r$$

$$w = \sqrt{\frac{g}{r}} = 0.057 \text{ rad/s}$$

The required rotation rate of the cylinder is 1 rev/110 s □
(For a description of proposed cities in space, see the article by Gerard K. O'Neill in *Physics Today,* Sept. 1974.)

12.37 **PROBLEM:** In introductory physics laboratories, a typical Cavendish balance for measuring the gravitational constant *G* uses lead spheres of masses 1.5 kg and 15 g whose centers are separated by about 4.5 cm. Calculate the gravitational force between these spheres, treating each sphere as a point mass at the center of the sphere.

SOLUTION:

$$F = \frac{G m^1 m^2}{r^2} = \frac{\left(6.67 \times 10^{-11} \text{ N} \cdot \text{m}^2\right)(1.5 \text{ kg})(0.015 \text{ kg})}{\text{kg}^2 \left(4.5 \times 10^{-2} \text{ m}\right)^2}$$

$$F = 7.41 \times 10^{-10} \text{ N} = 741 \text{ pN} \quad □$$

12.45 **PROBLEM:** X-ray pulses from Cygnus X-1, a celestial x-ray source, have been recorded during high-altitude rocket flights. The signals can be interpreted as originating when a blob of ionized matter orbits a black hole with a period of 5 ms. If the blob were in a circular orbit about a black hole whose mass was 20 times the mass of the Sun, what would be the radius of the orbit?

SOLUTION: The gravity outside a black hole is ordinary gravity. For the orbiting blob $\Sigma F = ma$ becomes

$$\frac{GM_h M_b}{r^2} = \frac{M_b v^2}{r} = \frac{M_b}{r}\left(\frac{2\pi r}{T}\right)^2$$

So we have Kepler's third law

$$T^2 GM_h = 4\pi^2 r^3$$

$$r = \left(\frac{T^2 GM_h}{4\pi^2}\right)^{1/2}$$

$$r = \left(\frac{\left(5 \times 10^{-3}\ \text{s}\right)^2 \left(6.67 \times 10^{-11}\ \text{N}\cdot\text{m}^2\right)(20)\left(1.99 \times 10^{30}\ \text{kg}\right)}{\left(4\pi^2\right)\left(\text{kg}^2\right)}\right)^{1/3}$$

$$r = 119\ \text{km}\ \square$$

12.49 **PROBLEM:** Two stars of mass M and m, separated by a distance of d, revolve in circular orbits about their center of mass (Fig. 12.49). Show that each star has a period given by

$$T^2 = \frac{4\pi^2}{G(M+m)}d^3$$

(*Hint:* Apply Newton's second law to each star, and note that the center-of-mass condition requires that $Mr_2 = mr_1$, where $r_1 + r_2 = d$.)

SOLUTION: For the star of mass M and orbital radius r_2,

$$\Sigma F = ma \text{ reads } \frac{GMm}{d^2} = \frac{Mv_2^2}{r_2} = \frac{M}{r_2}\left(\frac{2\pi r_2}{T}\right)^2$$

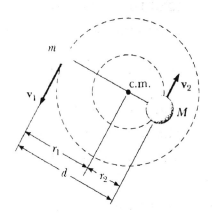

For the other star,

$$\frac{GMm}{d^2} = \frac{mv_1^2}{r_1} = \frac{m}{r_1}\left(\frac{2\pi r_1}{T}\right)^2$$

Clearing of fractions, we then obtain simultaneous equations

Figure 12.49

$$GmT^2 = 4\pi^2d^2r_2$$

$$GMT^2 = 4\pi^2d^2r_1$$

Adding lets us eliminate r_1 and r_2:

$$G(M+m)T^2 = 4\pi^2d^2(r_1 + r_2) = 4\pi^2d^3$$

$$T^2 = \frac{4\pi^2d^3}{G(M+m)}$$

In a visual binary star system T, d, r_1, and r_2 can be measured, so the mass of each component can be computed.

CHAPTER 13

QUESTIONS

6   **QUESTION:** Estimate the force of the atmosphere on a person's chest in view of the fact that atmospheric pressure is about 10^5 Pa.

ANSWER: For an order-of-magnitude estimate, think of your chest as a square 40 cm on each side. Then the air outside will exert a perpendicular force of $F = PA \cong 10^5 \text{ Pa} \left(\dfrac{1 \text{ N}}{1 \text{ Pa} \cdot \text{m}^2} \right) (0.4 \text{ m})^2 \sim 10^4 \text{ N}$

The air in your lungs, the blood in your arteries and veins, and the protoplasm in each cell exert nearly the same pressure, so the wall of your chest is in equilibrium.

9   **QUESTION:** If a helium-filled balloon is placed in a freezer, will its volume increase, decrease, or remain the same?

ANSWER: The rubber in a typical balloon is easy to stretch. The helium or air inside is nearly at atmospheric pressure. As the temperature drops the volume will decrease slowly. The rubber wall moves in slowly, just maintaining equality of pressure outside and inside. So in $PV = nRT$, V and T decrease by the same factor while P, n, and of course R stay constant. The process is an isobaric cooling or isobaric contraction.

10   **QUESTION:**   What happens to a helium-filled rubber balloon released
      into the air? Does it expand or contract? Does it stop rising at
      some height?

      ANSWER: Imagine the balloon rising into uniform-temperature air.
      The air cannot be uniform in pressure because the lower layers sup-
      port the weight of all the air above them. The rubber in a typical
      balloon is easy to stretch and stretches or contracts until inter-
      ior and exterior pressures are nearly equal. So as the balloon
      rises _it expands_; we are considering an isothermal expansion with P
      decreasing as V increases by the same factor in $PV = nRT$. If the
      rubber wall is very strong it will eventually contain the helium at
      higher pressure than the air outside but at the same density, so
      that the balloon will stop rising. More likely, the rubber will
      stretch and break, releasing the helium to keep rising and "boil
      out" of the Earth's atmosphere.

18   **QUESTION:**   When the metal ring and metal sphere in Figure 13.13 are
      both at room temperature, the sphere does not fit through the ring.
      After the ring is heated, the sphere can pass through the ring.
      Why does this occur?

      ANSWER: The thermal expansion of the ring is not like the
      inflation of a blook-pressure cuff or the swelling of a sore
      throat. Rather it is like a photographic enlargement. Every
      linear dimension, including the hole diameter, increases by the
      same factor. The atoms around the circumference of the hole push
      each other further away. A metal rim can be fitted tightly to a
      wheel by putting it on when the rim is hot.

CHAPTER 13

PROBLEMS

13.1 **PROBLEM:** Use the facts that normal atmospheric pressure is 1.013×10^5 Pa and the Earth's radius is 6.37×10^6 m to estimate the mass of the Earth's atmosphere.

SOLUTION: The pressure of the atmosphere is the weight of the air above each unit area on the Earth's surface. The atmosphere is very shallow compared to the Earth's radius, so the acceleration of gravity is nearly uniform at 9.80 m/s^2 over its height. Then

$$P = \frac{F}{A} = \frac{\text{weight of atmosphere}}{4\pi \, r^2} = \frac{mg}{4\pi \, r^2}$$

$$m = \frac{P4\pi \, r^2}{g}$$

$$m = \frac{(1.013 \times 10^5 \text{ Pa})\,4\pi(6.37 \times 10^6 \text{ m})^2}{9.80 \text{ m/s}^2}\left(\frac{N}{Pa \cdot m^2}\right)\left(\frac{kg \cdot m}{N \cdot s^2}\right)$$

$$m = 5.27 \times 10^{18} \text{ kg} \ \square$$

13.9   **PROBLEM:**   Liquid nitrogen has a boiling point of -195.81°C at atmospheric pressure. Express this temperature in (a) degrees Fahrenheit, and (b) kelvin.

SOLUTION:

(a)   This temperature is below the ice point of water by 195.81 Celsius degrees. One hundred Celsius degrees represent the same temperature change as 212°F - 32°F = 180F°, so boiling liquid N_2 is below freezing water by

$$195.81 \ C° \left(\frac{180 \ F°}{100 \ C°} \right) = \frac{9}{5} (195.81 \ C°) = 352 \ F°$$

Thus its temperature is 32°F - 352 F° = -320°F □

(b)   The ice point is 273.15 K, so liquid N_2 boils at 273.15 K - 195.81 K = 77.3 K. Note that a kelvin of temperature change is the same size as a celsius degree.

13.15   **PROBLEM:**   A copper telephone wire is strung, with little sag, between two poles that are 35 m apart. How much longer is the wire on a summer day, with T_C = 35°C, than on a winter day, with T_C = -20°C?

SOLUTION: The change in length between cold and hot conditions is

$$\Delta L = \alpha L_0 \Delta T = 17 \times 10^{-6} (C°)^{-1} \ (35 \ m) [35°C - (-20°C)]$$

$$\Delta L = 3.27 \ cm \ □$$

13.23 **PROBLEM:** The active element of a certain laser is made of a glass rod 30 cm long by 1.5 cm in diameter. If the temperature of the rod increases by 65°C, find the increase in rod's (a) length, (b) diameter, and (c) volume. ($\alpha = 9 \times 10^{-6}(°C)^{-1}$)

SOLUTION:

(a) $\Delta L = \alpha L_0 \Delta T = 9 \times 10^{-6}(C°)^{-1}(0.30 \text{ m})(65 \text{ C°}) = 1.76 \times 10^{-4} \text{ m}$ ☐

(b)   The diameter is a linear dimension, so the same equation applies:

$\Delta L = \alpha L_0 \Delta T = 9 \times 10^{-6}(C°)^{-1}(0.015 \text{ m})(65 \text{ C°}) = 8.78 \times 10^{-6} \text{ m}$ ☐

(c)   The original volume of the cylinder is

$$\pi r^2 h = \pi \left(\frac{0.015 \text{ m}}{2}\right)^2 (0.30 \text{ m}) = 5.30 \times 10^{-5} \text{ m}^3$$

Then $\Delta V = \beta V_0 \Delta T = 3\alpha V_0 \Delta T$

$$\Delta V = 3(9 \times 10^{-6})(C°)^{-1}(5.30 \times 10^{-5} \text{ m}^3)(65 \text{ C°})$$

$$\Delta V = 9.30 \times 10^{-8} \text{ m}^3 = 93.0 \text{ mm}^3 \text{ ☐}$$

We could also compute this from the result of parts (a) and (b):

$$\Delta V = V_f - V_i = \frac{\pi}{4} d_f^2 h_f - \frac{\pi}{4} d_i^2 h_i$$

$$\Delta V = \frac{\pi}{4} \left[(d_i + \Delta d)^2 (h_i + \Delta h) - d_i^2 h_i\right]$$

$$\Delta V = \frac{\pi}{4} \left[d_i^2 h_i + 2 d_i \Delta d \, h_i + d_i^2 \Delta h + \begin{array}{c}\text{negligibly}\\ \text{small terms}\end{array} - d_i^2 h_i\right]$$

$$\Delta V = \frac{\pi}{4} [2 \times 0.015 \text{ m} \times 8.78 \times 10^{-6} \text{ m} \times 0.3 \text{ m}$$

$$+ 0.015 \text{ m}^2 \times 176 \times 10^{-6} \text{ m}]$$

$$\Delta V = 9.3 \times 10^{-8} \text{ m}^3$$

13.29   **PROBLEM:**   The mass of a hot air balloon and its cargo (not including the air inside) is 200 kg. The air outside is at a temperature of 10°C and a pressure of 1 atm = 1.013 x 10^5 Pa. The volume of the balloon is 400 m^3. To what temperature must the air in the balloon be heated before the balloon will lift off? (Air density at 10°C is 1.25 kg/m^3.)

SOLUTION: Consider 400 m^3 of air at 10°C. We call this the air displaced by the balloon. Its mass is

$m = \rho V$ = (1.25 kg)(400 m^3) = 500 kg and its weight is

$w = mg$ = (500 kg)(9.8 m/s^2) = 4900 N. This chunk of cold air would stand in mechanical equilibrium surrounded by other air at 10°C, so the air around it must exert on it an upward force, called the buoyant force, equal to 4900 N. (The surrounding air can do this because the air below is at somewhat higher pressure than the air above.) The cold air in which it is immersed will exert this same force, 4900 N up, on the balloon. The envelope and cargo weigh (200 kg)(9.8 m/s^2) = 1960 N, so the hot air inside can weigh no more than 4900 N - 1960 N = 2940 N. It can have a mass no larger than $m = w/g$ = (2940 N)/(9.8 m/s^2) = 300 kg. Think of dry air as 80% N_2 and 20% O_2 with a little bit of heavier argon and CO_2. Then its molar mass is 0.80(28.0 g/mole) + 0.20(32.0 g/mole) = 29 g/mole.
The quantity of hot air in the balloon can be at most
$n = m/M$ = 300 kg(10^3g/kg)/(29 g/mole) = 1.03 x 10^4 mole.
The balloon is open to the outside air, so the pressure inside will be equal to that outside, 1.013 x 10^5 Pa, and we have for the temperature $PV = nRT$.

$$T = \frac{PV}{nR} = \frac{\left(1.013 \times 10^5 \text{ N/m}^2\right)\left(400 \text{ m}^3\right) \text{mole} \cdot \text{K}}{(1.03 \times 10^4 \text{ mole})(8.314 \text{ J})} \left(\frac{\text{J}}{\text{N} \cdot \text{m}}\right)$$

T = 471 K $\square$ = 198°C

13.31 **PROBLEM:** An automobile tire is inflated with air that is originally at 10°C and normal atmospheric pressure. During the process, the air is compressed to 28% of its original volume and the temperature is increased to 40°C. What is the tire pressure? After the car is driven at high speed, the tire air temperature rises to 85°C and the interior volume of the tire increases by 2%. What is the new tire pressure? Express each answer in pascals.

SOLUTION: Let $P_iV_i = nRT_i$ represent the state of the air before it is pumped into the tire, and $P_fV_f = nRT_f$ represent its state after this compression.

Then, $\dfrac{P_fV_f}{P_iV_i} = \dfrac{T_f}{T_i}$

and $P_f = P_i\dfrac{V_i\,T_f}{V_f\,T_i} = (1.013 \times 10^5 \text{ Pa})\left(\dfrac{V_i}{0.28V_i}\right)\left(\dfrac{(273 + 40)\text{ K}}{(273 + 10)\text{ K}}\right)$

$P_f = 4.00 \times 10^5$ Pa $\square$

A tire pressure gage would read "gauge pressure", the excess over atmospheric pressure, namely 400 kPa - 101.3 kPa = 299 kPa. Now let $P_hV_h = nRT_h$ represent the state of the air in the hot tire. We have

$\dfrac{P_hV_h}{P_fV_f} = \dfrac{T_h}{T_f}$

$P_h = P_f\left(\dfrac{V_f}{V_h}\right)\left(\dfrac{T_h}{T_f}\right) = (4.00 \times 10^5 \text{ Pa})\left(\dfrac{V_f\,(358\text{ K})}{1.02V_f\,(313\text{ K})}\right) = 4.49 \times 10^5$ Pa $\square$

The gauge pressure corresponding to this absolute pressure is 348 kPa.

13.39 **PROBLEM:** A sealed cubical container 20 cm on a side contains three times Avogadro's number of molecules at a temperature of 20°C. Find the force exerted by the gas on one of the walls of the container.

SOLUTION: Having three times Avogadro's number of molecules means there are three moles:

$$n = \frac{N}{N_A} = \frac{3\,N_A}{N_A} = 3$$

The volume is given as $V = (0.2 \text{ m})^3 = 0.008 \text{ m}^3$.

So we can find the pressure:

$PV = nRT$

$$P = \frac{nRT}{V} = \left(\frac{(3 \text{ mole})\,(8.314 \text{ J})\,(293 \text{ K})}{(0.008 \text{ m}^3)(\text{mole} \cdot \text{K})} \right) \left(\frac{\text{N} \cdot \text{m}}{\text{J}} \right) = 9.14 \times 10^5 \text{ N/m}^2$$

Then the force on each wall is outward

$F = PA = (9.14 \times 10^5 \text{ N/m}^2)(0.2 \text{ m})^2 = 36.5 \text{ kN}$ $\square$

Observe that the total force the gas exerts on all six walls is zero, since force is a vector. More precisely, if we take account of pressure differences between top and bottom, the total force is the weight of the gas. If it were, say, nitrogen, its weight would be

$w = mg = nMg = (3 \text{ moles})(28 \times 10^{-3} \text{ kg/mole})(9.8 \text{ m/s}^2) = 0.8 \text{ N}$,

quite negligible compared to 36.5 kN.

13.41   **PROBLEM:**   A cylinder contains a mixture of helium and argon gas in equilibrium at a temperature of 150°C.   (a) What is the average kinetic energy of each type of molecule?   (b) What is the rms speed of each type of molecule?

SOLUTION:

(a)   Both kinds of molecules have the same average kinetic energy, sharing it in collisions:

$$\frac{1}{2} mv^2 = \frac{3}{2} k_B T = \frac{3}{2}(1.38 \times 10^{-23} \text{ J / K})(273 + 150)\text{K}$$

$$K = 8.76 \times 10^{-21} \text{ J } \square$$

(b)   From the periodic table, helium atoms have mass

$$4 \text{ u} = (4 \text{ u})\left(\frac{1.66 \times 10^{-27} \text{ kg}}{1 \text{ u}}\right) = 6.64 \times 10^{-27} \text{ kg.}$$

Equivalently, we can say the molar mass of helium is 4.00 g/mole, so the mass of one molecule is

$$\frac{M}{N_A} = \frac{4 \times 10^{-3} \text{ kg/ mole}}{6.02 \times 10^{23} \text{ molecules/ mole}} = 6.64 \times 10^{-27} \text{ kg}$$

Then $\tfrac{1}{2} mv^2 = 8.76 \times 10^{-21}$ J gives

$$v_{\text{rms}} = \left(\overline{v^2}\right)^{1/2} = \left(\frac{2(8.76 \times 10^{-21} \text{ J})}{6.64 \times 10^{-27} \text{ kg}}\left(\frac{\text{kg} \cdot \text{m}^2}{\text{J} \cdot \text{s}^2}\right)\right)^{1/2} = 1.62 \text{ km/s } \square$$

For argon $M = 39.9$ g/mole, so

$$m = \frac{39.9 \times 10^{-3} \text{ kg/ mole}}{6.02 \times 10^{23} \text{ molecule/ mole}} = 6.63 \times 10^{-26} \text{ kg}$$

$$v_{\text{rms}} = \left(\frac{2(8.76 \times 10^{-21} \text{ J})}{6.63 \times 10^{-26} \text{ kg}}\right)^{1/2} = 514 \text{ m/s } \square$$

13.53 **PROBLEM:** A fluid has a density ρ. (a) Show that the *fractional* change in density for the change in temperature ΔT is given by $\Delta\rho/\rho = -\beta\Delta T$. What does the negative sign signify? (b) Fresh water has a miximum density of 1.000 g/cm^3 at 4°C. At 10°C, its density is 0.9997 g/cm^3. What is β for water over this temperature interval?

SOLUTION:

(a) Consider a fixed-mass sample. Its density changes because its volume changes with temperature, according to
$\Delta V = \beta V_0 \Delta T$. In $\rho = m/V = mV^{-1}$ take the derivative of ρ with respect to V: $\dfrac{d\rho}{dV} = m(-1)V^{-2}$

So for small changes very nearly

$$\frac{\Delta\rho}{\Delta V} = -\frac{m}{V^2}$$

$$\Delta\rho = -\frac{m}{V^2}\Delta V$$

Divide by density to find the fractional change

$$\frac{\Delta\rho}{\rho} = -\frac{mV}{V^2 m}\Delta V = -\frac{\Delta V}{V} = -\beta\Delta T$$

The negative sign means that ρ drops as T increases.

(b) Take changes from 4°C to 10°C

$$\frac{0.9997 \text{ g/cm}^3 - 1.000 \text{ g/cm}^3}{0.9997 \text{ g/cm}^3} = -\beta(10°C - 4°C)$$

$$\beta = \frac{-3 \times 10^{-4}}{-6 \text{ C}°} = 5.00 \times 10^{-5}/\text{C}° \quad \square$$

13.57 **PROBLEM:** A vertical cylinder of cross-sectional area A is fitted with a tight-fitting, friction-less piston of mass m (Fig.13.57). (a) If there are n moles of an ideal gas in the cylinder at a temperature of T, determine the height, h, at which the piston will be in equilibrium under its own weight. (b) What is the value for h if $n = 0.2$ mol, $T = 400$ K, $A = 0.008$ m^2, and $m = 20$ kg?

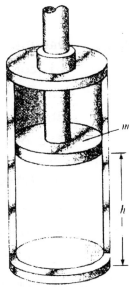

Figure 13.57

SOLUTION:

(a)   We suppose that air at pressure P_{atm} is above the piston. For the piston's equilibrium

$\Sigma F_y = ma_y$ is $-P_{atm}A - mg + PA = 0$, where P is the pressure exerted by the gas contained, given by $PV = nRT$ where $V = Ah$. We think of n, T, m, g, A, and P_{atm} as given, so we eliminate P and V by substitution and solve for h:

$$P = \frac{nRT}{Ah}$$

$$P_{atm}A - mg + \frac{nRT}{Ah}A = 0$$

$$h = \frac{nRT}{P_{atm}A + mg} \quad \square$$

(b) $$h = \frac{(0.2 \text{ mole})(8.314 \text{ J})(400 \text{ K})}{\left(1.013 \times 10^5 \text{ N/ m}^2\right)\left(0.008 \text{ m}^2\right) + (20 \text{ kg})\left(9.8 \text{ m/ s}^2\right)}$$

$$h = \frac{665 \text{ N} \cdot \text{m}}{810 \text{ N} + 196 \text{ N}} = 0.661 \text{ m} \quad \square$$

13.61   **PROBLEM:**   A hollow aluminum cylinder is to be fitted over a steel piston.   At 20°C the inside diameter of the cylinder is 99% of the outside diameter of the piston.   To what common temperature should the two pieces be heated in order that the cylinder just fit over the piston?

SOLUTION: Aluminum and steel both expand, but aluminum expands more, so heating can produce the fit.   Inner and outer diameters are both linear dimensions, so $\Delta L = \alpha L_0 \Delta T$ applies to both.

At 20°C we have $L_{Al} = 0.99 L_S$

At temperature T we require

$L_{Al} + \alpha_{Al} L_{Al} (T - 20°C) = L_S + \alpha_S L_S (T - 20°C)$

Substituting,

$0.99 L_S + \alpha_{Al} 0.99 L_S (T - 20°C) = L_S + \alpha_S L_S (T - 20°C)$

$(\alpha_{Al} 0.99 - \alpha_S)(T - 20°C) = 1 - 0.99$

$$T = 20° C + \frac{0.01}{0.99 \left(24 \times 10^{-6} \right) / C° - \left(11 \times 10^{-6} \right) / C°} = 804°C \quad \square$$

CHAPTER 14

QUESTIONS

3   **QUESTION:** Use the first law of thermodynamics to explain why the total energy of an isolated system is always constant.

ANSWER: An isolated system is an object or set of objects that exchanges neither work nor heat with its surroundings: the rest of the universe does no work on it and transfers no thermal energy to it, as the system goes from some initial to some final state. Then the first law says $\Delta U = 0$. Its initial is equal to its final internal energy in total amount, while energy may change from one form to another or move from one object to another in the system. Consider the bullet stopping in the pine board of Example 14.3. Take the bullet plus board as the system, initial point just before contact, and final point when macroscopic motion ceases. The wall does negative work on the bullet, but the outside world does no work on either, and not enough time elapses for any significant heat to transfer. So the total energy is constant. As mechanical energy disappears, it turns into an equal amount of extra internal energy. The organized motion of the bullet, all its molecules having the same velocity, is degraded into disorganized motion, random molecular vibration.

7   **QUESTION:**   What is wrong with the statement "Given any two
      bodies, the one with the higher temperture contains more thermal
      energy"?

      ANSWER: If body A has larger mass than body B, or if it has
      higher specific heat, or both, then it can contain more thermal
      energy than B even at a lower temperature.

11   **QUESTION:**   The air temperature above coastal areas is profoundly
      influenced by the large specific heat of water. One reason is
      that the thermal energy released during the cooling of 1 cubic
      meter of water by 1°C raises the temperature of an enormously
      larger volume of air by 1°C. Estimate this volume of air. (The
      specific heat of air is approximately 1.0 kJ/kg · °C. Take the
      density of air to be 1.3 kg/m^3.)

      ANSWER: The thermal energy released by one cubic meter of water,
      cooling by one Celsius degree, is

      $Q = mc\Delta T = (1000 \text{ kg})(4186)(\text{J/kg} \cdot \text{C°})(1 \text{ C°}) = 4.186 \text{ MJ}.$

      This heat will raise by 1 C° the temperature of air in mass

      $m = Q/(c\Delta T) = 4.186 \text{ MJ}/1(\text{kJ/kg} \cdot \text{C°})(1 \text{ C°}) = 4186 \text{ kg},$

      and in volume

      $$V = \frac{m}{\rho} = \frac{4186 \text{ kg}}{1.3 \text{ kg/m}^3} = 3.2 \times 10^3 \text{ m}^3.$$

CHAPTER 14

PROBLEMS

14.5   **PROBLEM:** The temperature of a silver bar rises by 10.0°C when it absorbs 1.23 kJ of thermal energy. The mass of the bar is 525 g. Determine the specific heat of silver.

SOLUTION:

$Q = mc\Delta T$

$$c = \frac{Q}{m\Delta T} = \frac{1.23 \text{ kJ}}{0.525 \text{ kg } 10 \text{ C}°} = 0.234 \text{ kJ/kg} \cdot \text{C}° \quad \square$$

14.7   **PROBLEM:** A 1.5-kg iron horseshoe initially at 600°C is dropped into a bucket containing 20 kg of water at 25°C. What is the final temperature of the horseshoe? (Neglect the heat capacity of the container.)

SOLUTION: As they come to a common final temperature T_f, horseshoe and the water exchange thermal energy, but we assume there is no heat flow through the bucket into the combined system:

$Q_{\text{into iron}} + Q_{\text{into water}} = 0$

$[mc(T_f - T_i)]_{\text{iron}} + [mc(T_f - T_i)]_{\text{water}} = 0$

$(1.5 \text{ kg})(448 \text{ J/kg} \cdot \text{C}°)(T_f - 600°\text{C})$

$\qquad\qquad + (20 \text{ kg})(4186 \text{ J/kg} \cdot \text{C}°)(T_f - 25°\text{C}) = 0$

$672 T_f \text{ J/C}° - 403 \text{ kJ} + 83720 T_f \text{ J/C}° - 2.09 \text{ MJ} = 0$

$84392 T_f \text{ J/C}° = 2.50 \text{ MJ}$

$T_f = 29.6°\text{C} \quad \square$

14.15 **PROBLEM:** A 3-g copper penny at 25°C drops a distance of 50 m
to the ground. (a) If 60% of the penny's inital potential
energy goes into increasing its internal energy, determine its
final temperature. (b) Does the result depend on the mass of
the penny? Explain.

SOLUTION: The initial energy is
$$U_g = mgy = (3 \times 10^{-3} \text{ kg})(9.8 \text{ m/s}^2)(50 \text{ m}) = 1.47 \text{ J}$$
We suppose that 40% of this energy is rubbed off by air and
ground, leaving in the coin 0.60 x 1.47 J = 0.882 J, which
becomes extra thermal energy. There is no thermal-energy-
transfer to the penny, but we imagine putting 0.882 J in with a
stove, to produce the same temperature change:

$$Q = mc\Delta T$$

$$\Delta T = \frac{Q}{mc} = \frac{0.882 \text{ J}}{(3 \times 10^{-3} \text{ kg})(387 \text{ J/ kg} \cdot \text{C}^\circ)} = 0.760 \text{ C}^\circ$$

So $T_f = T_i + \Delta T = 25°C + 0.760 \text{ C}^\circ = 25.8 \text{ °C}$ □

(b) The mass of the penny makes no difference.
Mathematically, it divides out like this:

$$\Delta T = \frac{Q}{mc} = \frac{0.60 \, U_g}{mc} = \frac{0.60 \, mgy}{mc} = 0.60 \, gy/c$$

14.25 **PROBLEM:** In an insulated vessel, 250 g of ice at 0°C is added to 600 g of water at 18°C. (a) What is the final temperature of the system? (b) How much ice remains?

SOLUTION: The ice starts to melt. If it all melts, it absorbs heat $Q = mL = (0.250 \text{ kg})(3.34 \times 10^5 \text{ J/kg}) = 83.5 \text{ kJ}$. The warmer water cools off. If it cools all the way to 0°C, it puts out heat $Q = mc\Delta T = (0.6 \text{ kg})(4186 \text{ J/kg} \cdot \text{C}°)18 \text{ C}° = 45.2 \text{ kJ}$ Since 83.5 kJ > 45.2 kJ, the warmer water cools all the way to 0°C $\square$, which is the final system temperature. The actual heat transferred is 45.2 kJ, to melt

$$m = \frac{Q}{L} = \frac{45.2 \times 10^3}{3.34 \times 10^5 \text{ J/ kg}} = 0.135 \text{ kg.}$$

So the ice remaining is

0.250 kg - 0.135 kg = 0.115 kg $\square$

14.29 **PROBLEM:** A 3-g lead bullet is traveling at a speed of 240 m/s when it embeds in a block of ice at 0°C. If all the thermal energy generated goes into melting ice, what quantity of ice is melted? (The heat of fusion for ice is 80 kcal/kg and the specific heat of lead is 0.03 kcal/kg · °C.)

SOLUTION: We suppose the bullet is originally at 0°C. Not all the ice melts, so the final temperature must be 0°C. The energy put into the ice is

$K = \frac{1}{2}mv^2 = \frac{1}{2}(3 \times 10^{-3} \text{ kg})(240 \text{ m/s})^2 = 86.4 \text{ J}$, to melt

$$m = \frac{Q}{L} = \frac{(86.4 \text{ J})(1 \text{ cal/ 4.186 J})}{80 \text{ cal/ g}} = 0.258 \text{ g} \square$$

14.33 **PROBLEM:** A sample of ideal gas is
expanded to twice its orginal volume of
1 m^3 in a quasi-static process for which
$P = \alpha V^2$, with $\alpha = 5.0$ atm/m^6, as shown in
Figure 14.33. How much work was done by
the expanding gas?

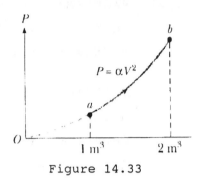

Figure 14.33

SOLUTION:

$$W = \int_i^f PdV = \int_{1m^3}^{2m^3}\left(5.0 \ \text{atm/m}^6\right)V^2 dV$$

$$W = \left(5.0 \ \text{atm/m}^6\right)V^3 / 3 \ \Big|_{1m^3}^{2m^3}$$

$W = (5.0 \ \text{atm/m}^6)(8 \ \text{m}^9 - 1 \ \text{m}^9)/3$

$W = (11.7 \ \text{atm} \cdot \text{m}^3)(1.013 \times 10^3 \ \text{N/m}^2 \cdot 1 \ \text{atm})(1 \ \text{J}/1 \ \text{N} \cdot \text{m})$

$W = 1.18$ MJ □

14.37 **PROBLEM:** A thermodynamic system undergoes a process in which
its internal energy decreased of 500 J. If at the same time,
220 J of work is done on the system, find the thermal energy
transeferred to or from the system.

SOLUTION:

$\Delta U = Q - W$

-500 J $= Q -$ (-200 J)

$Q = -720$ J

To solve the problem from the equation like this, we must
remember the sign conventions: ΔU is positive for an
increase, W is positive for work done by the system, and Q is
positive for heat input. So the answer is 720 J of thermal
energy <u>transferred out of the system</u> . □

14.41 **PROBLEM:** How much work is done by the steam when 1 mol of water at 100°C boils and becomes 1 mol of steam at 100°C and 1 atm pressure? Determine the change in the internal energy of the steam as it vaporizes. Consider the steam to be an ideal gas.

SOLUTION: The heat input is

$Q = mL = (0.018 \text{ kg})(2.26 \times 10^6 \text{ J/kg}) = 40.7 \text{ kJ}$

Of this, some energy comes out of the system as work done by expanding against the surrounding atmosphere, and the rest stays in the sample as increased internal energy. The original volume is

$$V = \frac{m}{\rho} = \frac{0.018 \text{ kg}}{1000 \text{ kg/ m}^3} = 18 \times 10^{-6} \text{ m}^3$$

We estimate the final volume from $PV = nRT$

$$V = \frac{nRT}{P} = \frac{(1 \text{ mole})(8.314 \text{ J})(373 \text{ K})(\text{m}^2)}{(\text{mole} \cdot \text{K})(1.013 \times 10^5 \text{ N})} = 3.06 \times 10^{-2} \text{ m}^3$$

So the water does work

$$W = \int_i^f P dv = P \int_i^f dv = P(V_f - V_i)$$

$W = (1.013 \times 10^5 \text{ N/m}^2)(3.06 \times 10^{-2} \text{ m}^3 - 18 \times 10^{-6} \text{ m}^3)$

$W = 3.10 \text{ kJ} \ \square$

Then $\Delta U = Q - W = 40.7 \text{ kJ} - 3.10 \text{ kJ} = 37.6 \text{ kJ} \ \square$

This problem is a good one to understand thoroughly. Our sign conventions for heat (input), work (output), and internal energy (increase) make all three positive in this process.

14.45 PROBLEM: Two moles of helium gas initially at a temperature of 300 K and a pressure of 0.4 atm is compressed isothermally to a pressure of 1.2 atm. Find (a) the final volume of the gas, (b) the work done by the gas, and (c) the thermal energy transferred. Consider the helium to behave as an ideal gas.

SOLUTION:

(a)   The initial volume comes from $P_i V_i = nRT_i$

$$V_i = \frac{nRT_i}{P_i}$$

$$V_i = \frac{(2 \text{ mole})(8.314 \text{ J/mole} \cdot \text{K})(300 \text{ K})}{0.4(1.013 \times 10^5 \text{ N/m}^2)}$$

$$V_i = 0.123 \text{ m}^3 \ \square$$

And the final volume is three times smaller:

$$V_f = \frac{nRT_f}{P_f}$$

$$V_f = \frac{(2 \text{ mole})(8.314 \text{ J/mole} \cdot \text{K})(300 \text{ K})}{1.2(1.013 \times 10^5 \text{ N/m}^2)}$$

$$V_f = 0.0410 \text{ m}^3 \ \square$$

(b)   In math class you learn to integrate $\int y \, dx$. The first step is to identify how y depends on x. Similarly, to calculate $W = \int_i^f P \, dv$ you must first identify how P depends on V. In this isothermal process, the constant-temperature both makes P inversely proportional to V:

$P = nRT/V$. So

$$W = \int_{V_I}^{V_f} nRTV^{-1} dV = nRT \ln V \Big|_{V_i}^{V_f} = nRT (\ln V_f - \ln V_i)$$

$$W = nRT \ln\left(\frac{V_f}{V_i}\right)$$

$$W = (2 \text{ moles}) \, 8.314 \, (\text{J/moleK}) \, 300 \text{ K} \ln (1/3) = -5.48 \text{ kJ} \ \square$$

14.45 (cont.)

    (c)   The ideal gas at constant temperature keeps constant internal energy. As it takes in 5.48 kJ of work from the falling piston, it must put out heat 5.48 kJ into the constant-temperature bath. This is described by

$$\Delta U = Q - W$$

$$Q = \Delta U + W = 0 - 5.48 \text{ kJ} = -5.48 \text{ kJ } \square$$

14.47 PROBLEM: A 1-kg block of aluminum is heated at atmospheric pressure so that its temperature increases from 22 °C to 40°C. Find (a) the work done by the aluminum, (b) the thermal energy added to the aluminum, and (c) the change in internal energy of the aluminum.

SOLUTION:

    (a)   The block does work by expanding against the constant pressure of the surrounding air:

$$W = \int_i^f Pdv = P\int_{V_i}^{V_f}dv = PV\Big|_{V_i}^{V_f} = P\left(V_f - V_i\right)$$

$$W = P\beta V\Delta T = P3\alpha(m/\rho)\ \Delta T$$

$$W = (1.013 \times 10^5 \text{ N/m}^2)3(24 \times 10^{-6}/C°)$$
$$(1 \text{ kg}/2.7 \times 10^3 \text{ kg/m}^3)(18 \text{ C°})$$

$$W = 48.6 \text{ mJ } \square$$

    (b) $Q = mc\Delta T = (1 \text{ kg})(900 \text{ J/kg} \cdot \text{C°})(18 \text{ C°}) = 16.2 \text{ kJ } \square$

    (c)   Kilojoules of heat go in as millijoules of work come out; the net input is the increase in internal energy:

$$\Delta U = Q - W = 16.2 \text{ kJ} - 48.6 \text{ mJ} = 16.2 \text{ kJ } \square$$

14.53   **PROBLEM:**   A water heater runs on solar power.   If the solar
collector has an area of 6 m^2, and the power per unit area
delivered by sunlight is 1000 W/m^2, how long does it take to
increase the temperature of 1 m^3 of water from 20°C to 60°C?

SOLUTION: Make sure you know that power describes how fast
energy is transferred.   The light _intensity_ is
I = 1000 W/m^2 = P/A. The _power_ of the heater is
P = IA = (1000 W/m^2)(6 m^2) = 6000 W = 6000 J/s
The _energy_ required to heat the water is
Q = $mc\Delta T$ = $\rho Vc\Delta T$ = (1000 kg/m^3)(1 m^3)(4186 J/kg · C°)(40 C°)
Q = 167 MJ
From P = E/t, the time required to accumulate this energy is
$$t = \frac{E}{P} = \frac{167 \times 10^6 \text{ J}}{6000 \text{ J/s}} = 27.9 \text{ ks } \square = 7.75 \text{ h}$$

14.55 **PROBLEM:** Around a crater formed by an iron meteorite. 75.0 kg of rock has melted under the impact of the meteorite. The rock has a specific heat of 0.8 kcal/kg · °C, a melting point of 500°C, and a latent heat of fusion of 48.0 kcal/kg. The origial temperature of the ground was 0.0°C. If the meteorite hit the ground with a terminal speed of 600 m/s, what was its minimum mass? Assume no energy loss to either the surrounding unmelted rock or the atmosphere during the impact. Disregard the heat capacity of the meteorite.

SOLUTION: The heat required to melt the rock is

$Q = mc\Delta T + mL$

$Q = (75 \text{ kg})(0.8 \text{ kcal/kg} \cdot \text{C}°)(500°C - 0°C) + (75 \text{ kg})(48 \text{ kcal/kg})$

$Q = 33.6 \times 10^6 \text{ cal}(4.186 \text{ J}/1 \text{ cal}) = 141 \times 10^6 \text{ J}$

We imagine all this energy carried in as kinetic energy of the meteorite:

$K = \tfrac{1}{2}mv^2 = 141 \times 10^6 \text{ J} = \tfrac{1}{2}m(600 \text{ m/s})^2$

$m = 781 \text{ kg}$ □

14.58 **PROBLEM:** A gas expands from a volume of 2 m^3 to a volume of 6 m^3 along two different paths, as shown in Figure 14.58. The thermal energy added to the gas along the path *IAF* is equal to 4 x 10^{5}cal. Find (a) the work done by the gas along the path *IAF*, (b) the work done along the path *IF*, (c) the change in internal energy of the gas, and (d) the thermal energy transferred in the process along the path *IF*.

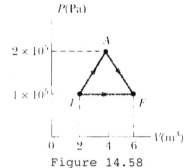

Figure 14.58

SOLUTION:

(a) The work the gas does in expanding along *IAF*,

$$W = \int_{\text{path IAF}} PdV$$ is the area between the path *IAF* and the V-axis. Since the segments *IA* and *AF* are straight, this is equal to the area of a rectangle of height 1.5 x 10^5 Pa and width (6 − 2) = 4 m^3:

$$W = (1.5 \times 10^5 \text{ Pa})(4 \text{ m}^3) = 600 \text{ kJ } \square$$

(b) $$W = \int_{IF} PdV = P\int_{I}^{F} dV = P\left(V_f - V_i\right)$$

$$W = (1 \times 10^5 \text{ Pa})(6 \text{ m}^3 - 2 \text{ m}^3) = 400 \text{ kJ } \square$$

(c) Consider the expansion along *IAF* and read the problem again. The increase in internal energy is the net energy input:

$$\Delta U = Q - W = 4 \times 10^5 \text{ cal}(4.186 \text{ J}/1 \text{ cal}) - 6 \times 10^5 \text{ J}$$

$$\Delta U = 1.07 \text{ MJ } \square$$

(d) Now along *IF* the change in internal energy must be the same because the initial and final states are the same:

$$\Delta U = 1.07 \text{ MJ and } W = 400 \text{ kJ so}$$

$$Q = \Delta U + W = 1.07 \text{ MJ} + 400 \text{ kJ} = 1.47 \text{ MJ} = 352 \text{ kcal } \square$$

14.69 **PROBLEM:** An ideal gas is enclosed in a cylinder with a movable piston on top of it. The piston has a mass of 8000 g and an area of 5 cm^2 and is free to slide up and down, keeping the pressure of the gas constant. How much work is done as the temperature of 0.2 mol of the gas is raised from 20°C to 300°C?

SOLUTION: METHOD ONE

The outside air is pushing down on the piston with force $P_{atm}A$. The sample gas pushes up with force PA. For the mechanical equilibrium of the piston,

$\Sigma F_y = 0$

$-P_{atm}A - mg + PA = 0$

$$P = P_{atm} + \frac{mg}{A} = 1.013 \times 10^5 \text{ N/m}^2 + \frac{(8 \text{ kg})(9.8 \text{ m/ s}^2)}{5 \times 10^{-4} \text{ m}^2} = 258 \text{ kPa}$$

The pressure stays constant as the volume increases from

$$V_i = \frac{nRT_i}{P} = \frac{(0.2 \text{ mole})(8.314 \text{ J/ mole} \cdot \text{K})(293 \text{ K})}{2.58 \times 10^5 \text{ Pa}}$$

$V_i = 1.888 \times 10^{-3} \text{ m}^3$

$$\text{to } V_f = \frac{nRT_f}{P} = \frac{(0.2 \text{ mole})(8.314 \text{ J/ mole} \cdot \text{K})(573 \text{ K})}{2.58 \times 10^5 \text{ Pa}}$$

$V_f = 3.692 \times 10^{-3} \text{ m}^3$

So the gas does work

$$W = \int_i^f P dv = P \int_i^f dv = PV \Big|_i^f = P(V_f - V_i)$$

$W = (258 \times 10^3 \text{ Pa})(3.692 - 1.888)(10^{-3} \text{ m}^3) = 466 \text{ J } \square$

METHOD TWO

It happens that we could get the answer without explicitly computing the pressure and volumes. For constant-pressure expansion of an ideal gas,

$W = P(V_f - V_i) = nRT_f - nRT_i$

$W = (0.2 \text{ mole})(8.314 \text{ J/mole} \cdot \text{K})(280 \text{ K}) = 466 \text{ J}$

CHAPTER 15

QUESTIONS

11 **QUESTION:** A living system, such as a tree, combines unorganized molecules (CO_2, H_2O), using sunlight, to produce leaves and branches. Is this reduction of entropy in the tree a violation of the second law of thermodynamics?

ANSWERS: An analogy due to Carnot is instructive: A waterfall continuously converts mechanical energy into thermal energy. It continuously creates entropy as the organized motion of the falling water turns into disorganized molecular motion. We humans put turbines into the waterfall, diverting some of the energy stream to our own use. Water falls spontaneously from higher to lower elevation and heat flows spontaneously from higher to lower temperature. Into the great flow of solar radiation from sun to Earth, living things put themselves. They live on energy flow. A basking snake diverts high-temperature heat through itself temporarily, before the energy is finally lost to low-temperature heat radiated into outer space. A tree builds organized cellulose molecules and we build libraries and babies who look like their grandmothers, all out of a thin diverted stream in the universal flow of energy crashing down to disorder. We do not violate the second law, for we build local reductions in the entropy of one thing within the inexorable increase in the total entropy of the universe. A growing tree or a human can do mechanical work, but cannot do better than the Carnot efficiency-limit on heat engines.

12   **QUESTION:** A diesel engine operates with a compression ratio of about 15. A gasoline engine has a compression ratio of about 6. Which engine runs hotter? Which engine (at least theoretically) can operate more efficiently?

ANSWERS: An internal-combustion engine takes in fuel and air at the environmental temperature and then compresses it. During the rapid (adiabatic) compression, work is put in. Heat has no time to come out, so the internal energy and temperature rise. The temperature-increase factor depends only on the nature of the molecules and the volume-decrease factor, called the compression ratio. With a larger compression ratio, a diesel engine will have its fuel and air hotter both before and after combustion, compared to a gasoline engine with spark plugs. Its Carnot efficiency limit is set by the higher temperature at which it takes in heat from the chemical energy in the fuel, so the diesel engine can have higher efficiency.

21   **QUESTION:**   If you shake a jar full of jelly beans of different
      sizes, the larger jelly beans tend to appear near the top, while
      the smaller ones tend to fall to the bottom.   Why does this
      occur?   Does the process violate the second law of
      thermodynamics?

      ANSWERS: Shaking opens up spaces between jelly-beans.   The
      smaller ones more often can fall down into spaces below them.
      The accumulation of larger candies on top and smaller ones on the
      bottom implies a small increase in order, a small decrease in one
      contribution to the total entropy, but the second law is not
      violated.   The total entropy increases as the system warms up,
      its increase in internal energy coming from the work put into
      shaking the box and also from a bit of gravitational energy loss
      as the beans settle compactly together.

      One analogous process of a bit of order appearing from disorder
      is "bathtub roulette".   Throw a slippery cake of soap into a wet
      bathtub and see how often it finally stops at the drain, the
      lowest point.   Another analogous process is life (see question
      11).

22 **QUESTION:** The divice in Figure 15.12, called a thermoelectric converter, is essentially a thermocouple driven by a temperature difference. If the left-hand photograph, both "legs" of the device are at the same temperature, and no electrical energy is produced. However, when one leg is at a higher temperature than the other, as in the right-had photograph, electrical energy is produced as the device extracts energy from the hot reservoir and drives a small electric motor. (a) Why does the temperature differential produce electrical energy in this demonstration? (b) In what sense does this intriguing experiment demonstrate the second law of thermodynamics?

ANSWERS: (a) A thermocouple is a pair of wires of different metals, with two junctions at their ends. When the junctions are at different temperatures, a small voltage appears around the loop, so that the device can be used to measure temperature or, here, to drive a small motor.

(b) The second law's statement that it is impossible for a cycling engine simply to extract heat from one reservoir, and convert it into work, directly describes the failure of the converter in the left-hand picture. To convert heat to work, the device must divert heat from the spontaneous flow from a hot to a cold reservoir, as in the right-hand picture. Mixing together the hot and the cold water leaves the total energy constant but increases the entropy as it equalizes the temperature, so there can be no energy flow in the left-hand picture.

CHAPTER 15

PROBLEMS

15.5   **PROBLEM:** A particular engine has a power output of 5 kW and an efficiency of 25%. If the engine expels 8000 J of thermal energy in each cycle, find (a) the heat absorbed in each cycle and (b) the time for each cycle.

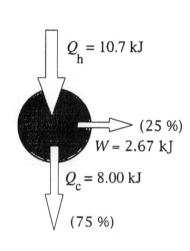

$Q_h = 10.7$ kJ

(25 %)
$W = 2.67$ kJ

$Q_c = 8.00$ kJ

(75 %)

Figure 15.5

SOLUTION: We know

$e = W/Q_h = 0.25$ and

$W = Q_h - Q_c = Q_h - 8000$ J/cycle,

so by substitution

$0.25 = (Q_h - 8000$ J$)/Q_h = 1 - 8000$ J$/Q_h$

8000 J$/Q_h = 0.75$

$Q_h = 8000$ J$/0.75 = 10.7$ kJ/cycle $\square$

and

$W = 10.7$ kJ $- 8$ kJ $= 2.67$ kJ/cycle

The power output is the useful work per time, so

$P = W/t = 5$ kW $= (2.67$ kJ$)/t$

$t = (2.67$ kJ/cycle$)/(5$ kJ/s$) = 0.533$ s/cycle $\square$

Looking at an energy flow diagram like Figure 15.1 is the best way to visualize how the process works.

15.11 **PROBLEM:** A steam engine is operated in a cold climate where the exhaust temperature is at 0°C. (a) Calculate the theoretical maximum efficiency of the engine using an intake steam temperature of 100°C. (b) If, instead, superheated steam at 200°C is used, find the engine's maximum possible efficiency.

SOLUTION:

(a) $e_c = 1 - T_c/T_h = 1 - 273$ K$/373$ K

$e_c = 0.268$ □

(b) $e_c = 1 - 273$ K$/473$ K $= 0.423$ □

Your most likely mistake is forgetting that the temperatures to which Carnot's equation refers are absolute temperatures.

15.13 **PROBLEM:** An ideal gas is taken through a Carnot cycle. The isothermal expansion occurs at 250°C, and the isothermal compres-sion takes place at 50°C. If the gas absorbs 1200 J of heat during the isothermal expansion, find (a) the heat expelled to the cold reservoir in each cycle and (b) the net work done by the gas in each cycle.

SOLUTION:

(a) For a Carnot cycle, $e_c = 1 - T_c/T_h$.
For any engine, $e = W/Q_h = 1 - Q_c/Q_h$. So for a Carnot engine $1 - Q_c/Q_h = 1 - T_c/T_h$ and
$Q_c = Q_h\ T_c/T_h = 1200$ J $\times$ 323 K$/523$ K $= 741$ J □

(b) $W = Q_h - Q_c = 1200$ J $- 741$ J $= 459$ J □

15.19   **PROBLEM:**   An electric generating plant has a power output of 500 MW.   The plant uses steam at 600°C and exhausts water at 40°C.   The system operates with one-half the maximum (Carnot) efficiency.   (a) At what rate is heat expelled to the environment?   (b) If the wast heat goes into a river whose flow rate is 1.2×10^6 kg/s, what is the rise in temperature of the river?

SOLUTION: (a) $e_c = 1 - T_c/T_h = 1 - 313$ K/873 K $= 0.641$, so the actual efficiency is $e = \frac{1}{2} \times 0.641 = 0.320$.   The power output is W/t, the time rate of useful work output.   We want to know Q_c/t, the rate of heat exhaust (thermal pollution).   We find it from $e = W/Q_h = (W/t)/(Q_h/t)$, so

$Q_h/t = (W/t)/e = 500$ MW/$0.320 = 1.56$ GW and $W = Q_h - Q_c$

$Q_c/t = Q_h/t - W/t = 1.56$ GW $- 500$ MW $= 1.06$ GW $\square$

(b)   Now $Q = mc\Delta T$ gives $(Q/t) = (m/t)c\Delta T$

$\Delta T = (Q/t)/(cm/t)$

$\Delta T = (1.06 \times 10^9 \text{ J/s})/[4186 \text{ (J/kg} \cdot \text{°C)} \times (1.2 \times 10^6 \text{ kg/s})]$

$\Delta T = 0.211$ °C

15.23   **PROBLEM:**   How much work is required, using an ideal Carnot refrigerator, to remove 1 J of thermal energy from helium gas at 4 K and reject this thermal energy to a room-temperature (293-K) environment?

SOLUTION: For a Carnot refrigerator the coefficient of performance is

$Q_c/W = T_c/(T_h - T_c)$

$W = Q_c(T_h - T_c)/T_c = (1 \text{ J})(293 - 4)\text{K}/4 \text{ K} = 72.2 \text{ J} \square$

15.25 **PROBLEM:** Calculate the change in entropy of 250 g of water when it is slowly heated from 20°C to 80°C.

(*Hint:* Note that $dQ = mc\, dT$.)

SOLUTION: To do the heating reversibly, put the water-pot successively into contact with reservoirs at temperatures 20°C + d, 20°C + 2d, . . . 80°C, where d is some small increment. Then

$$\Delta S = \int_{\text{reversible path}} \frac{dQ}{T} = \int_{T_i}^{f} mc\, \frac{dT}{T}$$

Here T means the absolute temperature. We would ordinarily think of dT as the change in the celsius temperature, but one celsius degree of temperature change is the same size as one kelvin of change, so dT is also the change in absolute $\underline{T}$. Then

$$\Delta S = mc \ln T \ \Big|_{T_I}^{T_F} = mc \ln \left(\frac{T_c}{T_i} \right)$$

$$\Delta S = (0.250 \ \text{kg})(4186 \ \text{J/kg·K}) \ln \frac{353 \ \text{K}}{293 \ \text{K}} = 195 \ \text{J/K} \ \square$$

15.29 **PROBLEM:** A 1500-kg car traveling at 20 m/s crashes into a concrete wall. If the air temperature is 20°C, what is the total entropy change?

SOLUTION: The original kinetic energy of the car, $K = \tfrac{1}{2}mv^2 = \tfrac{1}{2}(1500 \ \text{kg})(20 \ \text{m/s})^2 = 300 \ \text{kJ}$, becomes irreversibly 300 kJ of extra internal energy in car, wall, and then the surroundings. Since their total heat capacity is so large, their temperature ends up very nearly still at 20°C. To carry them reversibly to this same final state, imagine putting 300 kJ from a heater into the car, wall, and environment. Then

$$\Delta S = \int \frac{dQ}{T} = \frac{1}{T} \int dQ = \frac{Q}{T} = \frac{300 \ \text{kJ}}{293 \ \text{K}}$$

$$\Delta S = 1.02 \ \text{kJ/K} \ \square$$

15.31 **PROBLEM:** One mole of H^2 gas is
contained in the left side of the
container (Fig. 15.31; equal volumes
left and right). The right side is
evacuated. When the valve is opened,
hydrogen gas steams into the right

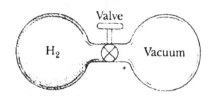

Figure 15.31

side. What is the final entropy change? Does the temperature
of the gas change?

SOLUTION: The expanding gas exerts no force against a moving
piston so it does no work. There is no time for heat to flow.
 Thus the internal energy must be constant. For an ideal gas V
is a function of temperature only, so the temperature stays
constant. To find the change in entropy in this irreversible
process, we consider letting the gas expand isothermally
against a piston to double in volume as heat enters it from a
constant-temperature both. Since
$dU = 0 = dQ - dW$,

$$dQ = dW = PdV = \frac{RT \, \frac{dV}{V}}{} \text{ and}$$

$$\Delta S = \int_i^f \frac{dQ}{T} = \int_{V_i}^{V_f} nR \frac{dV}{V} = nR \ln\left(\frac{V_f}{V_i}\right)$$

$$\Delta S = (1 \text{ mole})(8.314 \text{ J/mole} \cdot K) \ln 2 = 5.76 \text{ J/K } \square$$

15.37 **PROBLEM:** Repeat the procudure used to construct Table 13.1 (a) for the case in which you draw three marbles from your bag rather than four and (b) for the case in which you draw five rather than four.

SOLUTION:

(a)

Result	Possible Combinations	Total
all red	RRR	1
2R, 1G	RRG, RGR, GRR	3
1R, 2G	RGG, GRG, GGR	3
all green	GGG	1

(b)

Result	Possible Combinations		Total
all red	RRRRR		1
4R, 1G	RRRRG, RRRGR, RRGRR, RGRRR, GRRRR		5
3R, 2G	RRRGG, RRGRG, RGRRG, GRRRG, RRGGR,		
	RGRGR, GRRGR, RGGRR, GRGRR, GGRRR	10
2R, 3G	GGGRR, GGRGR, GRGGR, RGGGR, GGRRG,		
	GRGRG, RGGRG, GRRGG, RGRGG, RRGGG	10
1R, 4G	RGGGG, GRGGG, GGRGG, GGGRG, GGGGR		5
all green	GGGGG		1

15.43 **PROBLEM:** Every second at Niagara Falls, some 5000 m^3 of water falls a distance of 50 m. What is the increase in entropy per second due to the falling water? (Assume a 20°C environment).

SOLUTION: The original gravitational energy,

$Vg = mgy = \rho Vgy$

$Vg = (1000\ kg/m^3)(5000\ m^3)(9.8\ m/s^2)(50\ m) = 2.45\ GJ$

Irreversibly becomes extra internal energy in the water and its environment, which remains at 20°C because of its high heat capacity. To get to the same final state reversibly, we imagine putting 2.45 GJ of heat into the environment from a hotplate at nearly the same temperature. Then

$$\Delta S = \int_i^f \frac{dQ}{T} = \frac{1}{T}\int_i^f dQ = \frac{Q}{T} = \frac{2.45 \times 10^9\,J}{293\ K}$$

$\Delta S = 8.36\ MJ/K$ □

The rate of entropy increase is

$\Delta S/t = 8.36\ MJ/K \cdot s = 8.36\ MW/K$ □

15.45 **PROBLEM:** A house loses heat through the exterior walls and
roof at a rate of 5000 J/s = 5kW when the interior temperature
is 22°C and the outside temperature is -5°C. Calculate the
electric power required to maintain the interior temperature at
22°C for the following two cases. (a) The electric power is
used in electric resistance heaters (which convert all of the
electricity supplied into heat). (b) The electric power is
used to drive an electric motor that operates the compressor of
a heat pump (which has a coefficient of performance equal to
60% of the Carnot cycle value).

SOLUTION: $\dfrac{dQ}{dt} = 5000$ W $T_i = 295$ K $T_0 = 268$ K

(a)   If $\dfrac{\Delta Q}{\Delta t} = \dfrac{\Delta E}{\Delta t}$ then $P_{El} = 5000$ W $\square$

(b)   For a heat pump,

$$(COP)_{Carnot} = \frac{T_h}{\Delta T} = \frac{295}{27} = 10.92$$

Actual $COP = (0.6)(10.92) = 6.55 = Q_h/W = (Q_h/t)/(w/t)$

Therefore to bring 5000 W of heat into the house only
requires input power

$$\frac{w}{t} = \frac{(Q_h / t)}{COP} = \frac{5000 \text{ w}}{6.56} = 763 \text{ W } \square$$

15.46 **PROBLEM:** One mole of an ideal monatomic gas is taken through the cycle shown in Figure 15.46. The process AB is a rever-sible isothermal expansion. Calculate (a) the net work done by the gas, (b) the heat added to the gas, (c) the heat expelled by the gas, and (d) the efficiency of the cycle.

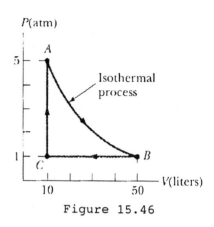

Figure 15.46

SOLUTION:

(a) The gas puts out work in process AB and takes in work in BC. We must find both to add up the net work. For the isothermal process AB, $W_{AB} = P_A V_A \ln\left(\frac{V_B}{V_A}\right)$

$$W_{AB} = (5)(1.013 \times 10^5 \text{ Pa})(10 \times 10^{-3} \text{ m}^3) \ln\left(\frac{50}{10}\right)$$

$$W_{AB} = 8.15 \times 10^3 \text{ J},$$

where we used 1 atm = 1.013×10^5 Pa and 1 L = 10^{-3} m^3

$$W_{BC} = P_B \Delta V = (1.01 \times 10^5 \text{ Pa})[(10 - 50) \times 10^{-3}] \text{ m}^3$$

$$W_{BC} = -4.05 \times 10^3 \text{ J}$$

$$W_{CA} = 0 \text{ and } W = W_{AB} + W_{BC} = 4.11 \times 10^3 \text{ J} \quad \square$$

15.46 (b) Since AB is an isothermal process, $\Delta U_{AB} = 0$, and

$Q_{AB} = W_{AB} = 8.15 \times 10^3$ J.

For an ideal monatomic gas $C_V = 3R/2$ and $C_P = 5R/2$.

$T_B = T_A = P_B V_B/(nR) = (1.01 \times 10^5)(50 \times 10^{-3})/R$

$T_B = 5.05 \times 10^3/R$.

Also, $T_C = P_C V_C/(nR) = (1.01 \times 10^5)(10 \times 10^{-3})/R$

$T_C = 1.01 \times 10^3/R$.

The gas takes in heat both in process AB and CA:

$$Q_{CA} = nC_V \Delta T = (1)\frac{3}{2} R \frac{(5.05 \times 10^3 - 1.01 \times 10^3)}{R} = 6.08 \times 10^3 \text{ J}$$

So the total heat absorbed is

$Q_{AB} + Q_{CA} = 8.15 \times 10^3$ J $+ 6.08 \times 10^3$ J $= 1.42 \times 10^4$ J $\square$

(c) The heat output we can find from

$Q_h = W + Q_C$

$Q_C = Q_h - W = 14.2$ kJ $= 4.11$ kJ $= 10.1$ kJ.

Alternatively, we can identify BC as the heat-exhaust process, with

$$Q_{BC} = nC_P \Delta T = \frac{5}{2}(1.01 \times 10^3)(1 - 5) = -1.01 \times 10^4 \text{ J } \square$$

(d) $e = \dfrac{W}{Q_h} = \dfrac{4.1 \times 10^3}{1.42 \times 10^4} = 0.289$ or 28.9% $\square$

15.47 **PROBLEM:** An athlete whose mass is 70 kg drinks 16 ounces (453.6 g) of refrigerated water. The water is at a temperature of 35°F. (a) Neglecting the temperature change of the body that results from the water intake (so that the body is re-garded as a reservoir at 98.6°F), find the entropy increase of the entire system. (b) Assume that the entire body is cooled by the drink and that the average specific heat of a human is equal to the specific heat of liquid water. Neglecting any other heat transfers and any metabolic heat release, find the athlete's temperature after she drinks the cold water, given an initial body temperature of 98.6°F. Under *these* assumptions, what is the entropy increase of the entire system? Compare this result with (a).

SOLUTION:

(a)   The body puts out heat, in amount

$$Q = mc\Delta T = (453.6 \text{ g})(4.186 \text{ J/g} \cdot °C)(98.6°F - 35°F)(5°C/9°F)$$

Q = 67.1 kJ at a temperature we here regard as constant at 98.6°F = 310 K. So its entropy change is

Q/T = 67.1 kJ/310 K = -216 J/K

That of the water is

$$\Delta S = \int_i^f \frac{dQ}{T} = \int_i^f mc \frac{dT}{T} = mc \ln T \Big|_{T_i}^{T_f} = mc \ln\left(\frac{T_f}{T_i}\right)$$

$$\Delta S = (453.6 \text{ g})\left(\frac{4.186 \text{ J}}{g \cdot °C}\right)\left(\frac{1 °C}{1 \text{ K}}\right)\ln\left(\frac{310 \text{ K}}{274.7 \text{ K}}\right)$$

ΔS = 230 J/K

So for the universe

ΔS = 230 J/K - 216 J/K = 13.4 J/K □

15.47 (b)   The cold drink mixes with the athlete according to

$$m_1 c_1 \Delta T = -m_2 c_2 \Delta T$$

$(-70{,}000 \text{ g})(4.186 \text{ J/g} \cdot °C)(T_f - 310.15 \text{ K})$

$\qquad = (453.6 \text{ g})(4.186 \text{ J/g} \cdot °C)(T_f - 274.82 \text{ K})$

$-293020 T_f + 90880153 \text{ K} = 1899 T_f - 521820 \text{ K}$

$91401972 \text{ K} = 294919 T_f$

$T_f = 309.92 \text{ K}$

Now we can use equation 15.11 to find the total entropy change:

$$\Delta S = m_1 c_1 \ln (T_f / T_1) + m_2 c_2 \ln (T_f / T_2)$$

$\Delta S = (70{,}000 \text{ g})(4.186 \text{ J/g} \cdot \text{K}) \ln (309.92/310.15)$

$\qquad + (453.6 \text{ g})(4.186 \text{ J/g} \cdot \text{K}) \ln (309.92/274.8 \text{ K})$

$\Delta S = -217.38 \text{ J/K} + 228.23 \text{ J/K} = 10.9 \text{ J/K} \ \square$

As the body temperature changes very little, so the different assumptions yield comparable entropy changes.

CHAPTER 16

QUESTIONS

4   **QUESTION:**   If a suspended object, A, is attracted to object B, which is charged, can we conclude that object A is charged? Explain.

ANSWER: Object A might have a charge opposite in sign to that of B, but it also might be a neutral conductor. In this latter case, the non-uniform field created by B would polarize A, pulling charge of one sign to the near face and pushing an equal amount of charge of the other sign to the far face, as in Figure 16.4(a). Then the force of attraction on the near side would be slightly larger than the force of repulsion on the far side, so the net force on A is toward B.

7   **QUESTION:**   When defining the electric field, why is it necessary to specify that the magnitude of the test charge be very small (i.e., $q \rightarrow 0$ taken as the limit)?

ANSWER: If the charges creating the field are on movable objects, or if they are on conductors, then a large test charge would exert a force back on them and make them move, changing the electric field at just the location where we are trying to measure it.

9   **QUESTION:** An uncharged, metallic-coated Ping-Pong ball is placed in the region between two vertical parallel metal plates. If the two plates are charged, one positive and one negative, describe the motion the Ping-Pong ball undergoes.

ANSWER: The two plates create a region of uniform electric field between them, pointing from the positive to the negative plate. If we were to do this experiment in the orbiting space shuttle, then an uncharged weightless ball could hang at rest in this region, feeling no total force. But if it is disturbed so as to touch one plate, say the negative one, it will immediately take up some of the negative charge and feel a force that will accelerate it to the positive plate. There it will dump its negative charge, acquire a positive charge, and start to accelerate back to the negative plate. The ball will rattle back and forth between the plates until it has ferried all their charges across to make both neutral.

13   **QUESTION:** A very large, thin, flat plate of aluminum of area A has a total charge of Q uniformly distributed over its surfaces. The same charge is spread uniformly over only the *upper* surface of an otherwise identical glass plate. Copmpare the electric fields just above the centers of the plates' upper surfaces.

ANSWER: The book's Equation 16.18 $E = \sigma_{conductor}/\epsilon_0$ for the field outside the aluminum looks different from Equation 16.17 $E = \sigma_{insulator}/2\epsilon_0$ for the field around the charged surface of the glass. But the aluminum carries charge on both of its faces so the density is $\sigma_{conductor} = Q/2A$, while this glass carries charge only on area A: $\sigma_{insulator} = Q/A$. Then the two fields are $Q/2A\epsilon_0$ the same in magnitude, and both are perpendicular to the plates.

15   **QUESTION:**   A uniform electric field exists in a region of space
      in which there are no charges.   What can you conclude about the
      net electric flux through a gaussian surface placed in this
      region of space?

      ANSWER: The net flux through any gaussian surface is zero.   We
      can argue it two ways.   Any surface contains zero charge so
      Gauss's law says the flux is zero.   The field is uniform, so the
      field lines entering one side of the closed surface come out the
      other side, and the net flux is zero.

22 **QUESTION:** Richard Feynman once said that if two persons stood at arm's length from each other and each person had 1% more electrons than protons, the force of repulsion between the two people would be enough to lift a "weight" equal to that of the entire Earth. Carry out an order-of-magnitude calculation to substantiate this assertion.

ANSWER: Suppose each person has mass 70 kg. In terms of elementary particles, each consists of precisely equal numbers of protons and electrons and a nearly equal number of neutrons. The electrons comprise very little of the mass, so we find the number of protons-and-neutrons in each person:

$$70 \text{ kg}\left(\frac{1 \text{ u}}{1.66 \times 10^{-27} \text{ kg}}\right) = 4 \times 10^{28} \text{ u}$$

Of these nearly one-half, 2×10^{28} are protons. One percent of this is 2×10^{26}, constituting a charge of

$$2 \times 10^{26}(1.6 \times 10^{-19} \text{ C}) = 3 \times 10^7 \text{ C}$$

Thus Feynman's force is

$$F = \frac{k \, q_1 q_2}{r^2} - \frac{\left(9 \times 10^9 \text{ N} \cdot \text{m}^2\right)\left(3 \times 10^7 \text{ C}\right)^2}{\left(c^2\right)(0.5 \text{ m})^2} = 4 \times 10^{25} \text{ N},$$

where we have used a half-meter arm's length. The mass of the Earth in a gravitational field of 9.8 m/s^2 would weigh

$$w = mg = (6 \times 10^{24} \text{ kg})(10 \text{ m/s}^2) = 6 \times 10^{25} \text{ N}$$

Thus the forces are of the same order of magnitude.

CHAPTER 16

PROBLEMS

16.5   **PROBLEM:** Calculate the magnitude and direction of the Coulomb force on each of the three charges in Figure 16.5

SOLUTION: The forces are as shown in the sketch.   The magnitudes of the forces are as follows:

$$F_1 = \frac{k_e q_1 q_2}{r_{12}^2} = (9 \times 10^9 \ N \cdot m^2/ \ C^2) \frac{(6 \times 10^{-6} \ C)(1.5 \times 10^{-6} \ C)}{(3 \times 10^{-2} \ m)^2} = 90 \ N$$

$$F_2 = \frac{k_e q_1 q_3}{r_{13}^2} = (9 \times 10^9 \ N \cdot m^2/ \ C^2) \frac{(6 \times 10^{-6} \ C)(2 \times 10^{-6} \ C)}{(5 \times 10^{-2} \ m)^2} = 43.2 \ N$$

$$F_3 = \frac{k_e q_2 q_3}{r_{23}^2} = (9 \times 10^9 \ N \cdot m^2/ \ C^2) \frac{(1.5 \times 10^{-6} C)(2 \times 10^{-6} C)}{(2 \times 10^{-2} m)^2} = 67.5 \ N$$

The net force on the 6 mC charge = $F_1 - F_2$ = 46.8 N ☐ (to the left).

The net force on the 1.5 mC charge = $F_1 + F_3$ = 157.N ☐ (toward right).

The net force on the -2 mC charge = $F_2 + F_3$ = 110.7 N ☐ (toward left).

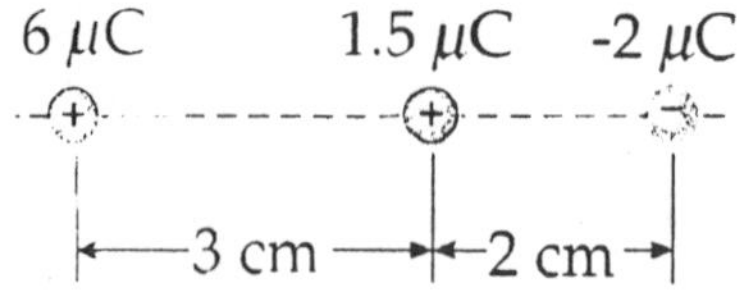

Figure 16.5

16.7 **PROBLEM:** Three charges are arranged as shown in Figure 16.7. Find the magnitude and direction of the electrostatic force on the charge at the origin.

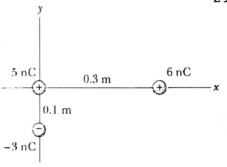

Figure 16.7

SOLUTION: We choose to find the field at the origin created by the other two charges:

$$E = \frac{k_e q}{r^2} \hat{r} + \frac{k_e q}{r^2} \hat{r}$$

$$E = \frac{(8.99 \times 10^9 \text{ N} \cdot \text{m}^2)(6 \times 10^{-9} \text{ C})}{c^2(0.3 \text{ m})^2} (-i) + \frac{8.99 \times 10^9 \text{ N} \cdot \text{m}^2}{c^2} \frac{(-3 \times 10^{-9} \text{ C})}{(0.1 \text{ m})^2} j$$

$E = 599 \text{ N/C}(-i) + 2700 \text{ N/C } (-j)$

$E = \sqrt{599^2 + 2700^2} \text{ N/C at Arctan } 2700/599 \text{ below the } -x \text{ axis}$

$E = 2.76 \text{ kN/C at } 77.5° \text{ below the } -x \text{ axis}$

$E = 2.76 \text{ kN/C at } 257°$

Now the force on the 5 nC is

$F = qE = (5 \times 10^{-9} \text{ C})(2.76 \times 10^3 \text{ N/C}) \text{ at } 257°$

$F = 13.8 \text{ } \mu\text{N at } 257° \ \square$

16.13   **PROBLEM:**   A continuous line of charge lies along the x axis, extending from $x = +x_0$ to $+\infty$.   This line carries a uniform linear charge density λ_0.   What are the magnitude and direction of the electric field at the origin?

   SOLUTION: A bit of the line, between x and $x + dx$, will have charge $\lambda_0 dx$ and will at the origin create electric field

$$d\mathbf{E} = \left(k_e \frac{dq}{r^2} \right)\hat{\mathbf{r}} = \left(k_e \lambda_0 \frac{dx}{x^2} \right)(-\mathbf{i})$$

   The total field is

$$\mathbf{E} = \int_{\text{all charge}} d\mathbf{E} = \int_{x_0}^{\infty} k_e \lambda_0 (-\mathbf{i}) x^{-2} dx$$

$$\mathbf{E} = -k_e \lambda_0 \mathbf{i} \frac{x^{-1}}{-1} \bigg|_{x_0}^{\infty} = (k_e \lambda_0 \mathbf{i}) \left[\frac{1}{\infty} - \frac{1}{x_0} \right]$$

$$\mathbf{E} = \left(k_e \frac{\lambda_0}{x_0} \right)(-\mathbf{i}) \ \square$$

16.15   **PROBLEM:**   Each of the electrons in a particle beam has a kinetic energy of 1.6×10^{-17} J.   What are the magnitude and direction of the electric field that will stop these electrons in a distance of 10 cm?

   SOLUTION: The required electric field will be in the direction of motion.   We know, work done $= \Delta K$ so, $-Fs = -\frac{1}{2} mv_0^2$ (since the final velocity $= 0$)   which becomes $eEs = \frac{1}{2} mv_0^2$ and

$$E = \frac{\frac{1}{2} mv_0^2}{es} = \frac{1.6 \times 10^{17} \text{ J}}{(1.6 \times 10^{19} \text{ C})(0.1 \text{ m})}$$

$$E = 10^3 \text{ N/C} \ \square \ \text{(in direction of electrons motion)}$$

16.19 **PROBLEM:** Three equal positive charges, q, are at the corners of an equilateral triangle with sides of length a, as in Figure 16.40. (a) At what points in the plane of the charges (other than ∞) is the electric field zero? (b) What are the magnitude and direction of the electric field at the point P due to the two charges at the base of the triangle?

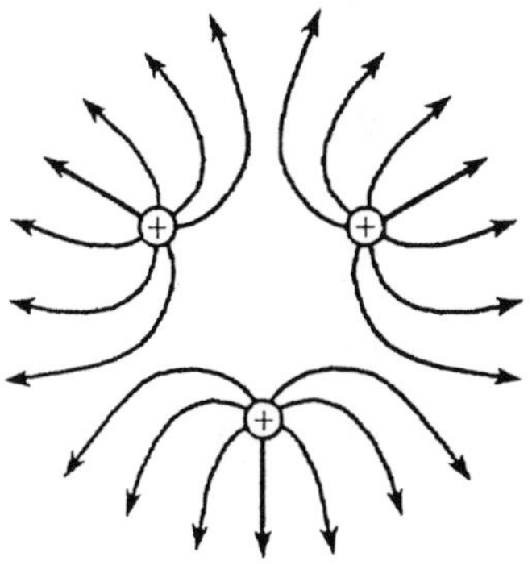

SOLUTION:

(a) The electric field has the general appearance shown. It is zero at the center, where the three charges indivifually produce fields that cancel out. [There are also three other points, but they are hard to find.}

Figure 16.19

(b) You will likely need to review vector addition in Chapter one.

$$\mathbf{E} = {}_e\frac{q\hat{\mathbf{r}}}{r^2} + k_e\frac{q\hat{\mathbf{r}}}{r^2}$$

$$\mathbf{E} = k_e\frac{q}{a^2} \quad \text{to the right and upward at } 60°$$

$$+ k_e\frac{q}{a^2} \quad \text{to the left and upward at } 60°.$$

The x components, positive and negative $k_eq \cos 60°/a^2$, add to zero, leaving

$$\mathbf{E} = k_eq \cos 60° \ \mathbf{j}/a^2 + k_eq \cos 60° \ \mathbf{j}/a^2$$

$$\mathbf{E} = 2k_eq \cos 60° \ \mathbf{j}/a^2 \ \square$$

The minus sign means that more lines enter the surface than leave it.

16.21 PROBLEM: A uniformly charged ring of radius 10 cm has a total charge of 75 μC. Find the electric field on the *axis* of the ring at (a) 1 cm, (b) 5 cm, (c) 30 cm, and (d) 100 cm from the center of the ring.

SOLUTION: Using the result of Example 16.5,

$$E = \frac{k_e x Q}{(x^2 + a^2)^{3/2}} = \frac{(9 \times 10^9)(75 \times 10^{-6})x}{(x^2 + 0.1^2)^{3/2}} = \frac{6.75 \times 10^5 \ x}{(x^2 + 0.01)^{3/2}}$$

(a) $x = 0.01$ m $\mathbf{E} = 6.65 \times 10^6 \mathbf{i}$ N/C $\square$

(b) $x = 0.05$ m $\mathbf{E} = 2.41 \times 10^7 \mathbf{i}$ N/C $\square$

(c) $x = 0.3$ m $\mathbf{E} = 6.40 \times 10^6 \mathbf{i}$ N/C $\square$

(d) $x = 1$ m $\mathbf{E} = 6.65 \times 10^5 \mathbf{i}$ N/C $\square$

16.29 PROBLEM: A 40-cm-diameter loop is rotated in a uniform electric field until the position of maximum electric flux is found. The flux in this position is measured to be 5.2×10^5 Nm^2/C. What is the electric field strength?

SOLUTION: In the orientation for maximum flux, the field is perpendicular to the area, so

$$\Phi = \mathbf{E} \cdot \mathbf{A} = E\pi r^2 \cos 0° = 5.2 \times 10^5 \ N \cdot m^2/C = E\pi(0.2 \ m)^2$$

$$E = 4.14 \times 10^6 \ N/C \ \square$$

16.37 PROBLEM: The following charges are located inside a submarine: +5 μC, -9 μC, +27 μC, and -84 μC. Calculate the net electric flux through the submarine hull. Compare the number of electric field lines leaving the submarine with the number entering it.

SOLUTION: The total charge within the closed surface is

+5 μC - 9 μC + 27 μC - 84 μC = -61 μC, so the total flux is

$\Phi = q/\epsilon_0 = (-61 \times 10^{-6}$ C$)/(8.85 \times 10^{-12}$ C^2/N $\cdot$ m$^2)$

$\Phi = -6.89 \times 10^6$ N $\cdot$ m^2/C $\square$

The minus sign means that more lines enter the surface than leave it.

16.41 PROBLEM: A point charge, Q, is located just above the center of the flat face of a hemisphere of radius R, as shown in Figure 16.41. (a) What is the electric flux through the curved surface of this hemisphere? (b) What is the electric flux through the flat face of this hemisphere?

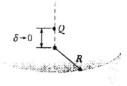

Figure 16.41

SOLUTION:

(a) With d very small, all points on the hemisphere are nearly at distance R from the charge, so the field everywhere on the curved surface is $k_e Q/R^2$ radially outward. Then the flux is this field strength times the area of half a sphere:

$$F_{curved} = \left(ke\frac{Q}{R^2}\right)\frac{1}{2}4\pi R^2 = \frac{1}{4\pi\,\varepsilon_0}Q\,2\pi = \frac{Q}{2\,\varepsilon_0} \square$$

(b) The whole closed surface encloses zero charge so Gauss's law says $F_{curved} + F_{flat} = 0$

$$F_{flat} = -F_{curved} = \frac{-Q}{2\,e_0} \square$$

16.43 PROBLEM: An insulating sphere of radius 10 mm has a uniform charge density of 6×10^{-3} C/m^3. Calculate the electric flux through a concentric spherical surface with the following radii: (a) $r = 5$ mm, (b) $r = 10$ mm, (c) $r = 25$ mm.

SOLUTION:

(a)   Inside radius 5 mm the charge is

$$q = \rho V = \rho \frac{4}{3} \pi r^3 = (6 \times 10^{-3} \ C/m^3)(4/3)\pi(5 \times 10^{-3} \ m)^3$$

$q = 3.14 \times 10^{-9}$ C.   So the flux is

$\Phi = q/\epsilon_0 = (3.14 \times 10^{-9} \ C)/(8.85 \times 10^{-12} \ C^2/N \cdot m^2)$

$\Phi = 355 \ N \cdot m^2/C \ \square$

(b)   and (c).   The total charge is
$(6 \times 10^{-3} \ C/m^3)(4/3)p(10^{-2} \ m)^3 = 2.51 \times 10^{-8}$ C.
So the flux through any surface enclosing it is

$\Phi = q/\epsilon_0 = (2.51 \times 10^{-8} \ C)/(8.85 \times 10^{-12} \ C^2/N \cdot m^2)$

$\Phi = 2.84 \ kN \cdot m^2/C \ \square$

1

16.45 **PROBLEM:** Consider a long cylindrical charge distribution of radius R with a uniform charge density of ρ. Find the electric field at a distance of r from the axis, where $r < R$.

SOLUTION: If ρ is positive, the field must everywhere be radially outward. Choose as the gaussian surface a cylinder of length L and radius r, contained inside the charged rod. Its volume is $\pi r^2 L$ and it encloses charge $\rho \pi r^2 L$. The circular end caps have no electric flux through them; there $\mathbf{E} \cdot d\mathbf{A} = 0$. The curved surface has $\mathbf{E} \cdot d\mathbf{A} = E dA \cos 0°$, and E must be the same strength everywhere over the curved surface.

Then $\int E \cdot dA = q / \epsilon_0$ becomes

$$E \int_{\text{curved surface}} dA = \frac{\rho \pi r^2 L}{\epsilon_0}$$

$$E 2\pi r L = \rho \pi r^2 L / \epsilon_0$$

Make sure you know that $2\pi r L$ is the lateral surface area of this cylinder. Solving,

$$E = \rho r / 2\epsilon_0$$

$\mathbf{E} = \rho r / 2\epsilon_0$ radially away from the axis $\square$

16.49   **PROBLEM:**   A square plate of copper with 50-cm sides is placed
        in the extended electric field of 8×10^4 N/C directed *perpen-*
        dicularly to the plate.   Find (a) the charge density of each
        face of the plate and (b) the total charge on each face.

        SOLUTION: The field, say in the x-direction, will push on and
        move free charges in the copper to polarize it: This deposits
        negative charge on the left face and an equal amount of
        positive charge on the right face.   This charge separation
        continues until there is zero field within the metal of the
        plate.   Take a gaussian surface as in Figure 16.34, to obtain
        from $\Phi = q/\epsilon_0$, $EA = \sigma A/\epsilon_0$.

        Then $\sigma = E\epsilon_0 = (8 \times 10^4$ N/C$)(8.85 \times 10^{-12}$ C^2/N $\cdot$ m$^2) = 708$ nC/m^2
        is the magnitude of the charge density on each face and
        $Q = \sigma A = (708$ nC/m$^2)(0.5$ m$)^2 = 177$ nC is the magnitude of the
        charge on each.

16.59 **PROBLEM:** A charged cork ball of mass 1 g is suspended on a light string in the presence of a uniform electric field, as shown in Figure 16.59. The electric field has components $E_x = 3 \times 10^5$ N/C and $E_y = 5 \times 10^5$ N/C. The ball is in equilibrium at q = 37°. Find (a) the charge on the ball and (b) the tension in the string.

SOLUTION:

(a) Let us sum force components to find

$$\Sigma F_x = E_x q - T \sin\theta = 0, \text{ and}$$

$$\Sigma F_y = E_y q + T \cos\theta - mg = 0$$

Combining these two equations, we get

$$q = \frac{mg}{(E_x \cot\theta + E_y)} = \frac{(0.001)(9.8)}{(3\cot 37° + 5 \times 10^5)} = 1.09 \times 10^{-8} \text{ C} \ \square$$

(b) From the two equations for ΣF_x and ΣF_y we also find

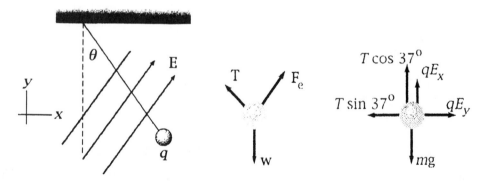

Figure 16.59

$$T = \frac{E \cdot q}{\sin 37°} = 5.43 \times 10^{-3} \text{ N} \ \square$$

16.61 **PROBLEM:** A solid, *insulating* sphere of radius a has a uniform charge density of r and a total charge of Q. Concentric with this sphere is an *uncharged, conducting* hollow sphere whose inner and outer radii are b and c, as in Figure 16.61. (a) Find the electric field in-tensity in the regions $r < a$, $a < r < b$, $b < r < c$, and $r > c$. (b) Determine the induced charge per unit area on the inner and outer surfaces of the hollow sphere.

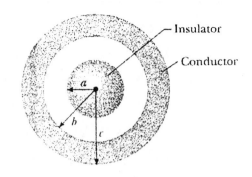

Figure 16.61

SOLUTION: (a) Choose as the gaussian surface a concentric sphere of radius r. The electric field will be everywhere perpendicular to its surface and will be uniform in strength over its surface.

The sphere of radius $r < a$ encloses charge $\rho \frac{4}{3} \pi r^3$, so

$\Phi = q/\epsilon_0$ becomes $E \cdot 4\pi r^2 = \rho \frac{4}{3} \pi r^3 / \epsilon_0$

$E = \rho r / 3\epsilon_0$ □.

For $a < r < b$ we have

$E4\pi r^2 = \frac{\rho 4}{3} \pi a^3 / \epsilon_0 = \frac{Q}{\epsilon_0}$

$E = \frac{\rho a^3}{3 \epsilon_0 r^2} = \frac{Q}{4\pi \epsilon_0 r^2}$ □

For $b < r < c$ we must have $E = 0$ □ because any non-zero field would be moving charges in the metal. Free charges did move in the metal to deposit charge $-Q_b$ on its inner surface, at radius b, leaving charge $+Q_c$ on its outer surface, at radius C. Since the shell as a whole is neutral, $Q_c - Q_b = 0$.

For $r > c$, $\Phi = q/\epsilon_0$ reads $E4\pi r^2 = (Q + Q_c - Q_b)/\epsilon_0$

$E = \frac{Q}{4\pi \epsilon_0 r^2}$ □

16.61 (cont.)

(b) For a gaussian surface of radius $b < r < c$ we have

$0 = (Q - Q_b)/\epsilon_0$ so $Q_b = Q$ and the charge density on the inner surface is $-Q_b/A = -Q/4\pi b^2$ □

Then $Q_c = Q_b = Q$ and the charge density on the outer surface is $+Q/4\pi c^2$ □

16.69 **PROBLEM:** Identical thin rods of length $2a$ carry equal charges, $+Q$, uniformly distributed along their lengths. The rods lie along the x axis with their centers separated by a distance of $b > 2a$ (Fig. 16.69). Show that the force exerted on the right rod is given

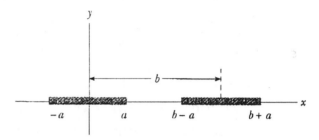

Figure 16.69

by $\quad F = \left(\dfrac{k_e Q^2}{4 a^2}\right) \ln\left(\dfrac{b^2}{b^2 - 4 a^2}\right)$

SOLUTION: Following Example 16.4, the left-hand rod creates electric field $E = \dfrac{k_e Q}{d\,(2a + d)}$. A segment of the right-hand rod has charge $dq = \lambda\,dx = Q\,\dfrac{dx}{2a}$, so it experiences a force

$dF = \dfrac{k_e QQ}{2a}\,\dfrac{dx}{d\,(d + 2a)}$. The total force on the rod is

$$F = \frac{k_e Q^2}{2a} \int_{x=b-2a}^{b} \frac{dx}{x\,(x + 2a)} = \frac{k_e Q^2}{2a}\left(-\frac{1}{2a}\ln\frac{2a + x}{x}\right)_{b-2a}^{b}$$

$$F = \frac{+ k_e Q^2}{4 a^2}\left(-\ln\frac{2a + b}{b} + \ln\frac{b}{b - 2a}\right) = \frac{k_e Q^2}{4 a^2}\ln\frac{b^2}{(b - 2a)\,(b + 2a)}$$

$$F = \left(\frac{k_e Q^2}{4 a^2}\right)\ln\left(\frac{b^2}{b^2 - 4 a^2}\right)$$

CHAPTER 17

QUESTIONS

4   **QUESTION:** If the electric potential at some point is zero, can you conclude that there are no charges in the vicinity of that point? Explain.

ANSWER: No! For a single charge, the potential it creates is zero only at infinite distance away, but if we have both positive and negative charges, there will be also plenty of points of zero potential among and around them. Look, for example, at the dipole in Figures 17.8(c) and 17.9 and in the computer plot on page 474. In a circuit on a lab bench we can declare any point to be ground, namely zero volts, by connecting it to the conducting earth, and the circuit will still function in the same way. The screen of a television picture tube, under continuous bombardment by electrons, is at zero volts.

6   **QUESTION:** Explain why, under static conditions, all points in a conductor must be at the same electric potential.

ANSWER: If two points on an object that can conduct electricity were at different potentials, then free charges in the object would move. (Free positive charges would migrate from higher- to lower-potential locations; free electrons would move quickly from negative to positive.) While the charges are moving the situation is not static, and the charges move until they have "discharged" all the high- and low-voltage areas, until the potential is everywhere equal.

15 **QUESTION:** If you want to increase the maximum operating voltage of a parallel-plate capacitor, describe how you can do this for a fixed plate separation.

ANSWER: Changing the area will change the capacitance and maximum charge, but not the maximum voltage. The question forbids us to increase the separation. We can put a material with higher dielectric strength between the plates or we could evacuate the space between the plates. We cannot really attain infinite electrical fields in vacuum, because atoms in the plates themselves will ionize, showing "thermionic emission" at very high fields.

18 **QUESTION:** Since the charges on the plates of a parallel-plate capacitor are equal and opposite, they attract each other. Hence, it takes positive work to increase the plate separation. What happens to the external work done in this process? (Assume the capacitor plates are removed from the charging battery.)

ANSWER: The work done in pulling the capacitor plates further apart becomes additional electric energy stored in the capacitor. The charge is constant and the capacitance decreases but the potential difference increases to drive up the energy. The electric field between the plates is constant in strength but fills more volume as you pull the plates apart.

23 **QUESTION:** The energy stored in a particular capacitor is increased fourfold. what is the accompanying change in (a) the charge? (b) the potential difference across the capacitor?

ANSWER: The charge Q and voltage V both double to make $U = CV^2/2 = Q^2/2C$ four times larger.

CHAPTER 17

PROBLEMS

17.9   **PROBLEM:**   (a) Calculate the speed of a proton that is accelerated from rest through a potential difference of 120 V. (b)   Calculate the speed of an electron that is accelerated through the same potential difference.

SOLUTION:

(a)   Energy is conserved as the proton moves from high to low potential; we take it as from +120 V to ground:

$K_i + U_i + \Delta K_{rc} = K_f + U_f$

(Reviw this work-energy theory of motion from Chapter 8 to the full extent necessary for you.)

$0 + qV + 0 = \frac{1}{2}mv^2 + 0$

$(1.6 \times 10^{-19} \text{ C})(120 \text{ V})\left(\dfrac{1 \text{ J}}{1 \text{ V} \cdot \text{C}}\right) = \frac{1}{2}(1.67 \times 10^{-27} \text{ kg})v^2$

$v = 1.52 \times 10^5 \text{ m/s}$ □

(b)   The electron will gain speed in moving the other way, from $V_i = 0$ to $V_f = 120$ V:

$K_i + U_i + \Delta K_{nc} = K_f + U_f$

$0 + 0 + 0 = \frac{1}{2}mv^2 + qV$

$0 = \frac{1}{2}(9.11 \times 10^{-31} \text{ kg})v^2 - (1.6 \times 10^{-19} \text{ C})(120 \text{ J/C})$

Note how the negative charge of the electron means it can have positive kinetic energy with zero total energy.

$v = 6.49 \times 10^6 \text{ m/s}$ □

This is less than one-tenth the speed of light, so we need not use the relativistic kinetic energy formula.

17.11 **PROBLEM:**   (1) Through what potential difference would an electron need to accelerate in order to achieve a speed of 60% of the speed of light, starting from rest? The speed of light is 3×10^8 m/s. (b) Repeat your calculation for a proton.

SOLUTION: Compare this problem to 17.9 above. For speeds larger than one-tenth the speed of light, $\frac{1}{2}mv^2$ gives noticeably wrong answers for kinetic energy, so we use

$K = mc^2(1 \; / \; \sqrt{1 - v^2 / c^2} - 1)$

$K = (9.11 \times 10^{-3} \text{kg})(3 \times 10^8 \text{m/ s})^2(1 \; / \; \sqrt{1 - 0.6^2} - 1)$

$K = 2.05 \times 10^{-14}$ J

Energy is conserved during acceleration:

$K_i + U_i + \Delta K_{nc} = K_f + U_f$

$0 + qV_i + 0 = 2.05 \times 10^{-14}$ J $+ qV_f$

The change in potential is $V_f - V_i$:

$V_f - V_i = \dfrac{-2.05 \times 10^{-14} \text{ J}}{q} = \dfrac{-2.05 \times 10^{-14} \text{ J}}{-1.6 \times 10^{-19} \text{ C}} = +128$ kV $\square$

The positive answer means that the electron speeds up in moving toward higher potential.

(b) We could use the same equations or we can reason thus: The proton has mass 1836 times larger and positive instead of negative charge. So it needs 1836 times more energy from a potential difference 1836 times larger in magnitude which it traverses from high to low voltage.

$V_f - V_i = -1836 \times 128$ kV $= -235$ MV $\square$

17.15 **PROBLEM:** An electron moving parallel to the x axis has an initial speed of 3.7×10^6 m/s at the origin. Its speed is reduced to 1.4×10^5 m/s at the point $x = 2$ cm. Calculate the potential difference between the origin and the point $x = 2$ cm. Which point is at the higher potential?

SOLUTION: Use the work-energy theorem to equate the energy of the electron at $x = 0$ and at $x = 2$ cm. The unknown will be the difference in potential $V_2 - V_0$.

$$K_i + U_i + \Delta K_{nc} = K_f + U_f$$

$$\tfrac{1}{2} mv_i^2 + qV_0 + 0 = \tfrac{1}{2} mv_f^2 + qV_2$$

$$\tfrac{1}{2} m\left(v_i^2 - v_f^2\right) = q(V_2 - V_0)$$

$$V_2 - V_0 = \frac{m\,(v_i^2 - v_f^2)}{2q}$$

Note that the electron's charge is negative.

$$V_2 - V_0 = \frac{(9.11 \times 10^{-31}\ \text{kg})\left((3.7 \times 10^6\ \text{m/ s})^2 - (1.4 \times 10^5\ \text{m/ s})^2\right)}{2(-1.6 \times 10^{-19}\ \text{C})}$$

$$V_2 - V_0 = -38.9 \text{ V } \square$$

The negative sign means that the 2 cm location is lower in potential than the origin. A positive charge would slow in free flight toward higher voltage, but the negative electron slows as it moves into lower potential. The 2 cm distance was unnecessary information for this problem. If the field were uniform, we could find it from $\Delta V = -Ed$

$$E = \frac{-\Delta V}{d} = \frac{-(-38.9 \text{ V})}{2 \times 10^{-2}\ \text{m}} = +1.95 \text{ kN/C in the direction of motion.}$$

17.23 **PROBLEM:** At a distance of r from a point charge, q, the electrical potential is $V = 400$ V and the magnitude of the electric field is $E = 150$ N/C. Determine the value of q and r.

SOLUTION: We have $V = \dfrac{k_e q}{r}$ and $|\mathbf{E}| = \left| \dfrac{k_e q}{r^2} \hat{\mathbf{r}} \right| = \dfrac{k_e q}{r^2}$

as two equations in the two unknowns q and r.

We can solve by substitution: $q = \dfrac{Vr}{k_e}$

$$E = \frac{k_e}{r^2} \frac{Vr}{k_e} = \frac{V}{r}$$

$$r = \frac{V}{E} = \left(\frac{400 \text{ V}}{150 \text{ N/ C}} \right) \left(\frac{J}{V \cdot C} \right) \left(\frac{N \cdot m}{J} \right) = 2.67 \text{ m} \ \square$$

$$q = \frac{Vr}{k_e} = \frac{(400 \text{ V})(2.67 \text{ m}) \text{ C}^2}{8.99 \times 10^9 \text{ N} \cdot \text{m}^2} \left(\frac{J}{V \cdot C} \right) \left(\frac{N}{J \cdot m} \right) = 1.19 \times 10^{-7} \text{ C} \ \square$$

You should check the answers by showing that $k_e q / r^2$ works out to be 150 N/C.

17.25 **PROBLEM:** A +2.8-µC charge is located on

the *y* axis at *y* = +1.6 m, and a −4.6-µC

charge is located at the origin. Calculate

the net electric potential at the point

(0.4 m, 0).

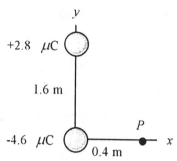

Figure 17.25

SOLUTION: Are you studying for an exam?

Then you should work out both the potential

at *P* and also the electric field. The potential is

$$V = \frac{k_e q}{r} + \frac{k_e q}{r}$$

$$V = \frac{(8.99 + 10^9\,N \cdot m^2)(2.8 \times 10^{-6}\,C)}{c^2\sqrt{(1.6\,m)^2 + (0.4\,m)^2}} + \frac{(8.99 \times 10^9\,N \cdot m^2)(-4.6 \times 10^{-6}\,C)}{c^2(0.4\,m)}$$

$V = 1.53 \times 10^4\,V - 10.3 \times 10^4\,V = -8.81 \times 10^4\,V$ □

The field is computed by vector addition, as in Chapter 16:

$$\mathbf{E} = \frac{k_e q}{r^2}\hat{r} + \frac{k_e q}{r^2}\hat{r}$$

$$\mathbf{E} = \frac{(8.99 \times 10^9\,N \cdot m^2)(2.8 \times 10^{-6}\,C)}{c^2\left[(1.6\,m)^2 + (0.4\,m)^2\right]} \quad \text{at Arctan}\left(\frac{1.6}{0.4}\right) \text{ below the } x\text{-axis}$$

$$+ \frac{(8.99 \times 10^9\,N \cdot C)(-4.6 \times 10^{-6}\,C)}{c^2(0.4\,m)^2} \quad \text{to the right}$$

$\mathbf{E} = 9.25 \times 10^3\,N/C$ at −76.0° + 2.58 × 10⁵ N/C ←

$\mathbf{E} = (9.25\,kN/C)\cos 76.0°\mathbf{i} + (9.25\,kN/C)\sin 76.0°(-\mathbf{j})$

$\qquad\qquad\qquad\qquad\qquad\qquad\qquad - (258\,kN/C)\mathbf{i}$

$\mathbf{E} = (2.24\,kN/C)\mathbf{i} - (8.98\,kN/C)\mathbf{j} - (258\,kN/C)\mathbf{i}$

$\mathbf{E} = (256\,kN/C)(-\mathbf{i}) + (8.98\,kN/C)(-\mathbf{j})$

$\mathbf{E} = 256\,kN/C$ at 182°

17.27 **PROBLEM:** The Bohr model of the hydrogen atom states that the electron can exist only in certain allowed orbits. The radius of each Bohr orbit is given by the expression $r = n^2(0.0529 \text{ nm})$ where $n = 1, 2, 3, \ldots$ Calculate the electric potential energy of a hydrogen atom when the electron (a) is in the first allowed orbit, $n = 1$; (b) is in the second allowed orbit, $n = 2$; (c) has escaped from the atom, $r = \infty$. Express your answers in electron volts.

SOLUTION: The electrical potential energy is given by

$$U = k_e \frac{q_1 q_2}{r}$$

(a) For the first allowed Bohr orbit,

$$U = \left(9 \times 10^9\right) \frac{\left(-1.6 \times 10^{-19}\right)\left(1.6 \times 10^{-19}\right)}{\left(0.0529 \times 10^{-9}\right)}$$

$$U = -4.37 \times 10^{-18} \text{ J} = \frac{-4.37 \times 10^{-18} \text{ J}}{1.6 \times 10^{-19} \text{ J/ eV}} = -27.3 \text{ eV} \ \square$$

(b) For the second allowed orbit,

$$U = \left(9 \times 10^9\right) \frac{\left(-1.6 \times 10^{-19}\right)\left(1.6 \times 10^{-19}\right)}{2^2\left(0.0529 \times 10^{-9}\right)}$$

$$U = -1.092 \times 10^{-18} \text{ J} = -6.81 \text{ eV} \ \square$$

(c) When the electron is at $r = \infty$,

$$U = \left(9 \times 10^9\right) \frac{\left(-1.6 \times 10^{-19}\right)\left(1.6 \times 10^{-19}\right)}{\infty} = 0 \ \square$$

17.28 **PROBLEM:** Show that the amount of work required to assemble four identical point charges of magnitude Q at the corners of a square with sides of length s is given by $5.41\ k_e \dfrac{Q^2}{s}$.

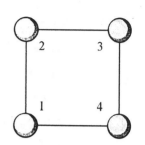

SOLUTION: Each charge creates potential on its own, and so injects energy into each other charge. We must add up $U = qV$ contributions for all pairs:

Figure 17.28

$$U = q_1 V_2 + q_1 V_3 + q_1 V_4 + q_2 V_3 + q_2 V_4 + q_3 V_4$$

$$U = \frac{q_1 k_e q_2}{r_{12}} + \frac{q_1 k_e q_3}{r_{13}} + \frac{q_1 k_e q_4}{r_{14}} + \frac{q_2 k_e q_3}{r_{23}} + \frac{q_2 k_e q_4}{r_{24}} + \frac{q_3 k_e q_4}{r_{34}}$$

$$U = \frac{Q k_e Q}{s} + \frac{Q k_e Q}{\sqrt{2}s} + \frac{Q k_e Q}{s} + \frac{Q k_e Q}{s} + \frac{Q k_e Q}{\sqrt{2}s} + \frac{Q k_e Q}{s}$$

$$U = \frac{k_e Q^2}{s}\left(4 + \frac{2}{\sqrt{2}}\right) = 5.414\ k_e \frac{Q^2}{s}$$

17.32 **PROBLEM:** Over a certain region of space, the electric potential is $V = 5x - 3x^2 y + 2yz^2$. Find the expressions for x, y, and z components of the electric field over this region. What is the magnitude of the field at the point P, which has coordinates (in meters) $(1,\ 0,\ -2)$?

SOLUTION: $V = 5x - 3x^2 y + 2yz^2$ \qquad Evaluate E at $(1, 0, -2)$

$$E_x = -\frac{\partial V}{\partial x} = -5 + 6xy = -5 + 6(1)(0) = -5$$

$$E_y = \frac{-\partial V}{\partial y} = +3x^2 - 2z^2 = 3(1)^2 - 2(-2)^2 = -5$$

$$E_z = \frac{-\partial V}{\partial z} = -4yz = -4(0)(-2) = 0$$

$$E = \sqrt{E_x^2 + E_y^2 + E_z^2} = \sqrt{-5^2 + (-5)^2 + 0^2} = 7.08 \text{ N/C} \quad \square$$

17.39 **PROBLEM:** A rod of length L (Fig. 17.39) lies along the x axis with its left end at the origin and has a *nonuniform* charge density of $\lambda = \alpha x$ (where a is a positive constant). (a) What are the units of the constant a? (b) Calculate the electric potential at point A, a distance of d from the left end of the rod.

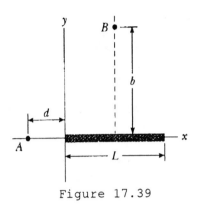

Figure 17.39

SOLUTION: (a) As a linear charge density, λ has units of coulomb per meter. Then $\alpha = \lambda/x$ has units C/m^2 □.

(b) Consider one bit of the rod at location x and of length dx. The amount of charge on it is $\lambda dx = \alpha x dx$. Its distance from A is $d + x$, so the bit of electric potential it creates at A is

$$dV = k_e \frac{dq}{r} = k_e \alpha x \frac{dx}{(d + x)}$$

We must integrate all these contributions for the whole rod, from $x = 0$ to $x = L$:

$$V = \int_{\text{all } q} dV = \int_{x=0}^{L} \frac{k_e \alpha x dx}{d + x}$$

To do the integral, change variables to $u = d + x$; $du = dx$; $u = d$ at $x = 0$ and $u = d + L$ at $x = L$:

$$V = \int_{u=d}^{d+L} \frac{k_e \alpha (u - d) du}{u} = k_e \alpha \int_{d}^{d+L} du - k_e \alpha d \int_{d}^{d+L} \left(\frac{1}{u}\right) du$$

17 **17.39 (cont)**

[Keep track of symbols: the <u>unknown</u> is V. The k_e, a, d, and L are <u>known</u> and constant. And x and u are variables, taking on all values over their ranges; they will not appear in the answer.]

$$V = k_e \alpha u \big|_d^{d+L} - k_e \alpha d \ln u \big|_d^{d+L}$$

$$V = k_e \alpha (d + L - d) - k_e \alpha d (\ln (d + L) - \ln d)$$

$$V = k_e \alpha L - k_e \alpha d \ln \left(\frac{d + L}{d} \right) \quad \square$$

We have the answer when the unknown is expressed in terms of the d, L, and α mentioned in the problem and the universal constant k_e.

17.45 **PROBLEM:** Is it feasible to construct a parallel-plate capacitor with its two plates separated by 0.10 mm and a capacitance of 1.0 F?

SOLUTION: $C = \dfrac{\kappa \in_0 A}{d}$

$$A = \frac{Cd}{\kappa \in_0} = (1 \quad F) \left(\frac{1 \text{ C}}{1 \text{ VF}} \right) \frac{0.1 \times 10^{-3} \text{ m N} \cdot \text{m}^2}{1 (8.85 \times 10^{-12} \text{ C}^2)} \frac{J}{N \cdot m} \frac{V \cdot C}{J}$$

$$A = 1.13 \times 10^7 \text{ m}^2 \quad \square$$

We could use two square sheets of aluminum foil, each 3.36 km on an edge. The cost would be like the cost of a large building.

17.53 **PROBLEM:** Find the charge on each of the capacitors in Figure 17.53.

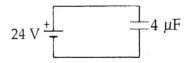

Figure 17.53

SOLUTION: We reduce the circuit in steps as shown below. Using the equivalent circuit, we find

$$Q_{total} = CV = (4 \ \mu C)(24 \ V) = 96 \ \mu C$$

Then in the second circuit,

$$V_{AB} = \frac{Q_{total}}{C_{AB}} = \frac{96 \ \mu C}{6 \ \mu C} = 16 \ V$$

$$V_{BC} = \frac{Q_{total}}{C_{BC}} = \frac{96 \ \mu C}{12 \ \mu C} = 8 \ V$$

Finally, using the first circuit above,

$$Q_1 = C_1 V_{AB} = (1 \ \mu F)(16 \ V) = 16 \ \mu C \ \square$$

$$Q_5 = C_5 V_{AB} = (5 \ \mu F)(16 \ V) = 80 \ \mu C \ \square$$

$$Q_8 = C_8 V_{BC} = (8 \ \mu F)(8 \ V) = 64 \ \mu C \ \square$$

$$Q_4 = C_4 V_{BC} = (4 \ \mu F)(8 \ V) = 32 \ \mu C \ \square$$

17.61 **PROBLEM:** A conducting slab with thickness d and area A is inserted into the space between the plates of a parallel-plate capacitor with spacing s and area A, as shown in Figure 17.61. What is the value of the capacitance of the system?

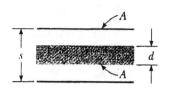

Figure 17.61

SOLUTION: If the capacitor is charged with charge Q, free charges will move across the slab to neutralize the electric field inside it, with the top and bottom faces of the slab then carrying charges $+Q$ and $-Q$. Then the capacitor with slab is electrically equivalent to two capacitors in series. Call x the upper gap, so $s-d-x$ is the distance between the lower two surfaces. The upper capacitor has $C_1 = \dfrac{\epsilon_0 A}{x}$ and the lower has

$$C_2 = \frac{\epsilon_0 A}{s - d - x}, \text{ so the combination has}$$

$$C = \frac{1}{1 / C_1 + 1 / C_2} = \frac{1}{x / \epsilon_0 A + (s - d - x) / \epsilon_0 A}$$

$$C = \frac{\epsilon_0 A}{s - d}$$

17.65 **PROBLEM:** A parallel-plate capacitor has plates of area 2 cm^2, a separation of 5 mm, and air between the plates. If a 12-V battery is connected to this capacitor, how much energy does it store?

SOLUTION: The capacitance is

$$C = \kappa \epsilon_0 A / d = 1(8.85 \times 10^{-12} \text{ C}^2/\text{N} \cdot \text{m}^2)(2 \times 10^{-4} \text{ m}^2)/(5 \times 10^{-3} \text{ m})$$

$$C = 3.54 \times 10^{-13} \text{ C}^2/\text{N} \cdot \text{m} = 3.54 \times 10^{-13} \text{ F}$$

The stored energy U is

$$U = CV^2/2 = (3.54 \times 10^{-13} \text{ C}^2/\text{J})(12 \text{ J/C})^2/2$$

$$U = 2.55 \times 10^{-11} \text{ J} \ \Box$$

17.67 **PROBLEM:** A parallel-plate capacitor has the charge Q and plates of area A. Show that the force exerted on each plate by the other is $F = Q^2/2\epsilon_0 A$. (*Hint:* Let $C = \epsilon_0 A/x$ for an arbitrary plate separation, x; then require that the work done in separating the two charged plates be $W = \int F\, dx$.)

SOLUTION: The electric field in the space between the plates is $E = \sigma/\epsilon_0 = Q/A\epsilon_0$, so you might think that the force on one plate is $F = QE = Q^2/A\epsilon_0$. But this is two times too large, because nothing can exert a force on itself. The force on one plate is exerted by the other, which creates electric field $E = \sigma/2\epsilon_0 = Q/2A\epsilon_0$, according to Example 16.10. Then the force on each plate is $F = Q_{\text{that plate}}\, E_{\text{the other plate}} = Q^2/2A\epsilon_0$. We can prove this by following the hint. The work done in separating the plates is the potential energy stored in the charged capacitor:

$$U = \frac{1}{2}\frac{Q^2}{C} = \int F\, dx \qquad \text{From the fundamental theorem of calculus,}$$

$dU = Fdx$, and

$$F = \frac{d}{dx}U = \frac{d}{dx}\frac{1}{2}\frac{Q^2}{C} = \frac{d}{dx}\frac{1}{2}\frac{Q^2}{\epsilon_0 A / x} = \frac{d}{dx}\frac{1}{2}\frac{Q^2 x}{\epsilon_0 A} = \frac{1}{2}\frac{Q^2}{\epsilon_0 A} \qquad \square$$

17.73 **PROBLEM:**   (a) How much charge can be placed on a capacitor with air between the plates before it breaks down, if the area of each of the plates is 5 cm^2?   (b) Find the maximum charge if polystyrene is used between the plates instead of air.

SOLUTION: The charge is $Q = CV$ with $C = \kappa\epsilon_0 A/d$ and with the electric field given by $E = V/d$.   So, $Q = \kappa\epsilon_0 Av/d = \kappa\epsilon_0 AE$.

(This is the same as saying $E = \dfrac{Q}{\kappa\,\epsilon_0\,A} = \dfrac{\sigma}{\kappa\,\epsilon_0}$, a generalized version of the result of Example 16.10.)   The maximum charge magnitude on each plate, before the dielectric lets a spark pass, is $Q_{max} = \kappa\epsilon_0 AE_{max}$.

(a)   For Air

$Q_{max} = 1(8.85 \times 10^{-12}\ c^2/N \cdot m^2)(5 \times 10^{-4}\ m^2)(3 \times 10^6\ N/C)$

$Q_{max} = 1.33 \times 10^{-9}\ C\ \Box$

(b)   For polystyrene

$Q_{max} = 2.56(8.85 \times 10^{12}\ c^2/N \cdot m^2)(5 \times 10^{-4}\ m^2)(24 \times 10^6\ N/C)$

$Q_{max} = 2.72 \times 10^{-7}\ C\ \Box$

17.81 **PROBLEM:** When two capacitors are connected in parallel, the equivalent capacitance is 4 μF. If the same capacitors are reconnected in series, the equivalent capacitance is one-fourth the capacitance of one of the two capacitors. Determine the two capacitances.

SOLUTION: For the parallel connection, $C_1 + C_2 = 4$ μF. For the series,

$$\frac{1}{C_1} + \frac{1}{C_2} = \frac{1}{C_1/4} = \frac{4}{C_1}, \text{ so } \frac{1}{C_2} = \frac{3}{C_1} \qquad C_1 = 3C_2.$$

We solve by substitution:

$3C_2 + C_2 = 4$ μF

$C_2 = 1.00$ μF □ and

$C_1 = 3C_2 = 3.00$ μF □

17.85 **PROBLEM:** A parallel-plate capacitor is constructed using a dielectric material whose dielectric constant is 3 and whose dielectric strength is 2×10^8 V/m. The desired capacitance is 0.25 μF, and the capacitor must withstand a maximum potential difference of 4000 V. Find the minimum area of the capacitor plates.

SOLUTION: $k = 3$, $E_{max} = 2 \times 10^8$ V/m $= V_{max}/d$ so $d = V_{max}/E_{max}$

For $C = \dfrac{\kappa \epsilon_0 A}{d} = 0.25 \times 10^{-6}$ F,

$$A = \frac{Cd}{k \epsilon_0} = \frac{CV_{max}}{\kappa \epsilon_0 E_{max}}$$

$$A = \frac{\left(0.25 \times 10^{-6}\right)(4000)}{3\left(8.85 \times 10^{-12}\right)\left(2 \times 10^8\right)} = 0.188 \text{ m}^2 \; \square$$

17.87 **PROBLEM:** The liquid-drop model of the nucleus suggests that high-energy oscillations of certain nuclei can split the nucleus into two unequal fragments plus a few neutrons. The fragments acquire kinetic energy from their mutual Coulombic repulsion. Calculate the Coulomb potential energy (in MeV) of two spherical fragments from a uranium nucleus having the following charges and radii: +38e and raidus 5.5×10^{-15} m; +54e and radius 6.2×10^{-15} m. Assume that the charge is distributed uniformly throughout the volume of each spherical fragment and that their surfaces are initially in contact at rest. (The electrons surrounding the nucleus can be neglected.)

SOLUTION: The problem is equivalent to finding the potential energy of a point charge +38e at distance 11.7×10^{-15} m from a point charge +54e:

$$U = qV = k_e \frac{q_1 q_2}{r_{12}}$$

$$U = \left(8.99 \times 10^9\right) \frac{(38)(54)(1.6 \times 10^{-19})^2}{(5.5 + 6.2) \times 10^{-15}}$$

$$U = 4.04 \times 10^{-11} \text{ J}$$

$$U = 253 \text{ MeV } \square$$

CHAPTER 18

QUESTIONS

10   **QUESTION:** If you were to design an electric heater using Nichrome wire as the heating element, what parameters of the wire could you vary to meet a specific power ouput, such as 1000 W?

   ANSWER: Having decided on the power desired (say 1000 W), choose the potential difference from which the device will draw energy (say 120 V). This determines the current ($I = P/V = 8.33$ A) and required resistance ($R = V/I = 14.4\Omega$). The material and operating temperature determine the resistivity of the heater, so what remains is to choose a length and a cross-sectional area for the resistor to satisfy $R = \rho\ell/A$. If you choose A very small, the filament will be fragile; if A is large, you will have to buy more material. Often ℓ and A must both be large enough for the heater to have enough surface area to lose heat fast enough to avoid too high a temperature.

11   **QUESTION:** Car batteries are often rated in ampere-hours. Does this designate the amount of current, power, energy, or charge that can be drawn from the battery?

   ANSWER: Because an ampere is a coulomb per second, an ampere-hour is 3600 C, the amount of <u>charge</u> that the battery can provide.

242

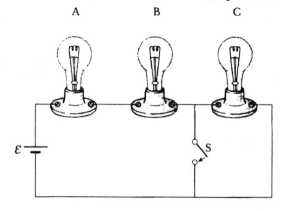

Figure 18.16

16 **QUESTION:** A series circuit
consists of three identical lamps
connected to a battery, as shown
in Figure 18.16. When switch S is
closed, (a) what happens to the
intensity of lamps A and B? (b)
What happens to the intensity of
lamp C? (c) What happens to the
current in the circuit? (d) Does the power dissipated in the
circuit increase, decrease, or remain the same?

ANSWER: Suppose $E = 12$ V and each lamp has $R = 2\ \Omega$. Before the
switch is closed the current is $(12\ \text{V})/(6\ \Omega) = 2\text{A}$. The power of
each lamp is $VI = I^2R = (2\ \text{A})^2 2\ \Omega = 8$ W, totalling
$12\ \text{V} \cdot 2\ \text{A} = 24$ W. Closing the switch makes it a zero-
resistance branch. All of the current through A and B will go
through the switch, and lamp C goes out. With less total resistance, the current in the battery $(12\ \text{V})/(4\ \Omega) = 3$ A will be
larger and lamps A and B will get brighter: $(3\ \text{A})2\ (2\ \Omega) = 18$ W,
totalling power $12\ \text{V} \cdot 3\ \text{A} = 36$ W, more than before.

CHAPTER 18

PROBLEMS

18.5 **PROBLEM:** A teapot with a surface area of 700 cm^2 is to be silver plated. It is attached to the negative electrode of an electrolytic cell containing silver nitrate ($Ag^+NO_3^-$). If the cell is powered by a 12.0-V battery and has a resistance of 1.8 Ω, how long does it take to build up a 0.133-mm layer of silver on the teapot? (Density of silver = 10.5 x 10^3 kg/m^3.)

SOLUTION: We look up the molar mass of silver,

107.9 g/mole = $\dfrac{107.9 \ kg}{kmole}$ and its density 10.5 x 10^3 kg/m^3.

(Each Ag+ ion has charge +1e.)

Volume = (area)(thickness) = (700 x 10^{-4} m^2)(0.133 x 10^{-3} m)

= 9.31 x 10^{-6} m^3

The mass of silver deposited = (density)(volume)

= (10.5 x 10^3 kg/m^3)(9.31 x 10^{-6} m^3) = 9.78 x 10^{-2} kg

and the number of silver atoms deposited = n

n = 9.78 x 10^{-2} kg $\dfrac{6.02 \ x \ 10^{26} \ atoms}{107.9 \ kg}$ = 5.454 x 10^{23}

$I = \dfrac{V}{R} = \dfrac{12 \ V}{1.8 \ \Omega}$ = 6.667 A = 6.667 C/s

and $\Delta t = \dfrac{\Delta Q}{I} = \dfrac{ne}{I} = \dfrac{\left(5.454 \ x \ 10^{23}\right)\left(1.6 \ x \ 10^{-19} \ C\right)}{6.667 \ C/ \ s}$ = 1.31 x 10^4 s

Δt = 3.64 h $\square$

18.11   **PROBLEM:**   A wire with a resistance of R is lengthened to 1.25 times its original length by pulling it through a small hole. Find the resistance of the wire after it is stretched.

SOLUTION: We assume its density and mass are unchanged, so its volume is constant. Call its original cross-section area A_1 and the area of the hole A_2. Then $\ell_1 A_1 = \ell_2 A_2$, $\ell_2 = 1.25\ \ell_1$ so $\ell_1 A_1 = 1.25\ \ell_1 A_2$, and $A_1 = 1.25\ A_2$. Then $R = r l_1 / A_1$ changes to

$$R_2 = \rho \ell_2 / A_2 = \rho\ 1.25\ \ell_1 / (A_1 / 1.25) = 1.25^2 \rho \ell_1 / A_1$$

$$R_2 = 1.25^2 R = 1.56 R \ \square$$

18.12   **PROBLEM:**   Suppose that you wish to fabricate a uniform wire out of 1 g of copper. If the wire is to have a resistance of $R = 0.5\ \Omega$, and all of the copper is to be used, what will be (a) the length and (b) the diameter of this wire?

SOLUTION: Don't mix up symbols! Call the density ρ_d and the resistivity ρ_r. Then from $\rho_d = m/v$, the volume is $v = A\ell = m/\rho_d$. The resistance is $R = \rho_r \ell / A$. We can solve for l by eliminating A:

$$A = m / \ell \rho_d$$

$$R = \rho_r \ell / (m / \ell \rho_d)$$

$$R = \rho_r \rho_d l^2 / m$$

$$\ell = \sqrt{\frac{mR}{\rho_r\,\rho_d}} = \sqrt{\frac{(10^{-3}\ \text{kg})\,(0.5\ \Omega)(m^3)}{(1.7 \times 10^{-8}\ \Omega \cdot m)\,(8.93 \times 10^3\ \text{kg})}} = 1.81\ \text{m}\ \square$$

To have a single diameter, the wire has a circular cross-section:

$$A = \pi r^2 = \pi (d/2)^2 = m / \ell r_d$$

$$d = \sqrt{4m\ /\ \pi \ell\ \rho_d} = \sqrt{4 \times 10^{-3}\ \text{kg}/ \left[\pi(1.81\ \text{m})(8.93 \times 10^3\ \text{kg}/m^3)\right]}$$

$$d = 0.280\ \text{mm}\ \square$$

18.17 **PROBLEM:** The electron beam emerging from a certain high-energy electron accelerator has a circular cross-section of radius 1 mm. (a) If the beam curretn is 8 μA, find the current density in the beam, assuming that it is uniform throughout. (b) The speed of the electrons is so close to the speed of light that it can be taken as $c = 3 \times 10^8$ m/s with negligible error. Find the electron density in the beam. (c) How long does it take for Avogadro's number of electrons to emergy from the accelerator?

SOLUTION: (a) $J = \dfrac{I}{A} = \dfrac{8 \times 10^{-6} \text{ A}}{\pi(1 \times 10^{-3} \text{ m})^2} = 2.55 \text{ A/m}^2$ □

(b) From $J = nev_d$, we have

$$n = \frac{J}{ev_d} = \frac{2.55 \text{ A/m}^2}{(1.60 \times 10^{-19} \text{ C})(3 \times 10^8 \text{ m/s})}$$

$n = 5.31 \times 10^{10} \text{ m}^{-3}$ □

(c) From $I = \Delta Q/\Delta t$, we have

$$\Delta t = \frac{\Delta Q}{I} = \frac{N_A e}{I} = \frac{(6.02 \times 10^{23})(1.60 \times 10^{-19} \text{ C})}{8 \times 10^{-6} \text{ A}}$$

$\Delta t = 1.20 \times 10^{10}$ s □ (or about 381 years!)

18.19 **PROBLEM:** An aluminum rod has a resistance of 1.234 Ω at 20°C. Calculate the resistance of the rod at 120°C by accounting for the changes in both the resistivity and the dimensions of the rod.

SOLUTION: For aluminum $\alpha_E = 3.9 \times 10^{-3}/°C$ (Table 18.1) is the temperature coefficient of resistivity, and $\alpha = 24 \times 10^{-6}/°C$ (Table 13.1) is the coefficient of thermal expansion. Now the new resistance is

$$R = \frac{\rho \ell}{A} = \frac{\rho_0 [1 + \alpha_E \Delta T] \ell (1 + \alpha \Delta T)}{A(1 + \alpha \Delta T)^2}$$

$$R = R_0 \frac{(1 + \alpha_E \Delta T)}{(1 + \alpha \Delta T)} = (1.234 \ \Omega) \frac{(1.39)}{(1.0024)}$$

$$R = 1.711 \ \Omega \ \square$$

18.27 **PROBLEM:** If the drift speed of free electrons in a copper wire is 7.84 $\times$ 10^{-4} m/s, calculate the electric field in the conductor.

SOLUTION: The electron density in copper, from Example 18.1, is 8.48 $\times$ 10^{28}/m^3. The current density in this wire is
$$J = nqv_d = (8.48 \times 10^{28}/m^3)(1.6 \times 10^{-19} \ C)(7.84 \times 10^{-4} \ m/s)$$
$$J = 1.06 \times 10^7 \ A/m^2$$
Now the microscopic form of Ohm's law is
$$J = \sigma E = E/\rho$$
$$E = \rho J = (1.7 \times 10^{-8} \ \Omega \cdot m)(1.06 \times 10^7 \ A/m^2)$$
$$E = 0.180 \ V/m \ \square$$

18.31 **PROBLEM:** Suppose that a voltage surge produces 140 V for a moment. By what percentage will the output of a 120-V, 100-W lightbulb increase, assuming the bulb's resistance does not change?

SOLUTION: We find the resistance:

$P_1 = V_1 I_1$

$I_1 = P_1/V_1 = (100 \text{ W})/(120 \text{ V}) = 833 \text{ mA}$

$R = V_1/I_1 = (120 \text{ V})/(0.833 \text{ A}) = 144 \text{ } \Omega.$

Now the current is larger,

$I_2 = V_2/R = (140 \text{ V})/(144 \text{ } \Omega) = 972 \text{ mA}$

and the power is a lot larger:

$P_2 = I_2 V_2 = (0.972 \text{ A})(140 \text{ V}) = 136 \text{ W}$

The percent increase is

$(136 \text{ W} - 100 \text{ W})/(100 \text{ W}) = 0.361 = 36.1\%$ □

18.43   **PROBLEM:**   Two 1.50-V batteries - with their positive terminals in the same direction - are inserted in series into the barrel of a flashlight.   One battery has an internal resistance of 0.255 Ω, the other an internal resistance of 0.153 Ω.   When the switch is closed, a current of 0.6 A occurs in the lamp.   (a) What is the lamp's resistance?   (b) What fraction of the power dissipated is dissipated in the batteries?

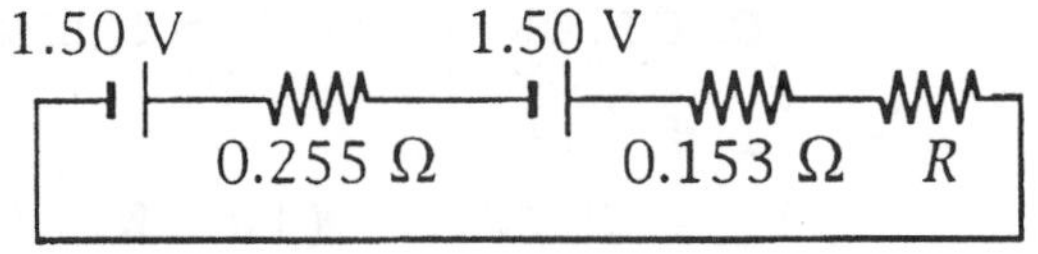

Figure 18.43

SOLUTION:

(a)   Kirchhoff's loop theorem says

$$1.5 \text{ V} - (0.255 \ \Omega)(0.6 \text{ A}) + 1.5 \text{ V}$$
$$- (0.153 \ \Omega)(0.6 \text{ A}) - R(0.6 \text{ A}) = 0$$
$$R = (2.76 \text{ V})/(0.6 \text{ A}) = 4.59 \ \Omega \ \square$$

(b)   The total power converted in the circuit is the power output of the emf's:

$$(1.5 \text{ V})(0.6 \text{ A}) + (1.5 \text{ V})(0.6 \text{ A}) = 1.8 \text{ W}$$

The power dissipated into heat in the batteries is

$$I^2R_1 + I^2R2 = (0.6 \text{ A})^2(0.255 \ \Omega + 0.153 \ \Omega) = 0.147 \text{ W}$$

So the fractional inefficiency is

$$(0.147 \text{ W})/(1.8 \text{ W}) = 0.0816 \ \square$$

18.45 **PROBLEM:** The current in a circuit is tripled by connecting a 500-Ω resistor in parallel with the resistance of the circuit. Determine the resistance of the circuit in the absence of the 500-Ω resistor.

SOLUTION: Before   After

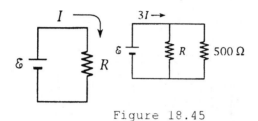

Figure 18.45

In both pictures, resistor R has the same voltage ε across it and so the same current I through it. Kirchhoff's junction rule says that the current in the 500 Ω is

$3I - I = 2I$. So $\varepsilon = 2I(500\ \Omega)$ and $\varepsilon = IR$ thus $IR = 2I(500\ \Omega)$ and $R = 1000\ \Omega$ $\square$

18.48   **PROBLEM:** Consider the circuit shown
        in Figure 18.48. Find (a) the
        current in the 20-Ω resistor and (b)
        the potential difference between
        points *a* and *b*.

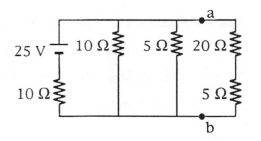

Figure 18.48(a) - Picture 1

SOLUTION: The given diagram is the same as this:

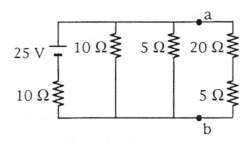

Figure 18.48(b) - Picture 2

The 20 W and 5 W are in series, so the equivalent is

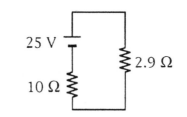

Figure 18.48(c) - Picture 3

Now the 10 W, 5 W, and 25 W resistors are in parallel.
Since 1/(1/10 + 1/5 + 1/25) = 2.94, the equivalent is

18.48 (cont)

At last we get

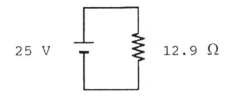

25 V   12.9 Ω

Figure 18.48   (d) - Picutre 4

Now we work backwards through the pictures, applying $I = V/R$ and
$V = IR$ alternately to every resistor, real and equivalent. The
12.9 Ω is connected across 25 V so the current through the 25 V
in every picture is $I = V/R = $ 25 V/12.9 Ω = 1.93 A.   In picture
(c), this 1.93 A goes through both 2.94 Ω and 10 Ω, to give the
voltage across them as $V = IR = $ (1.93 A)(2.94 Ω) = 5.68 V and
$V = IR = $ (1.93 A)(10 V) = 19.3 V.   In picture (b), the 5.68 V is
V_{ab} □, across each of the three parallel resistors.   The new
10 Ω carries current (5.68 V)/(10 Ω) = 0.568 A.   The 5 Ω carries
5.68 V/5 W = 1.14 A, and the 25 Ω carries $I = V/R = $ 5.68 V/25 Ω
= 0.227 A.   Now picture (a) being equivalent to picture (b) means
that this 0.227 A □ goes through both the 20 Ω and the new 5 Ω.
(The voltage across them are (20 Ω)(0.227 A) = 4.55 V and
(5 Ω)(0.227 A) = 1.14 V, adding to V_{ab} = 5.68 V, as computed
above.)

18.51   **PROBLEM:**   Determine the current in
each of the branches of the circuit
shown in Figure 18.51(a).

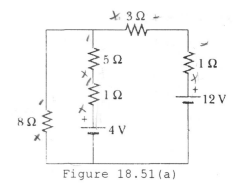

Figure 18.51(a)

SOLUTION: Suppose the current is
downward in the 8 Ω.   Call it I_1.
Suppose the current is downward in the
5 Ω, 1 Ω, and 4 V.   Call it I_2.
Suppose the current is upward in the 12 V.   Call
it I_3.   Then the current rule says $I_3 = I_1 + I_2$.
The voltage rule says about a clockwise trip
around the left-hand mesh, starting from the lower
left, $+I_1(8 \ \Omega) - I_2(5 \ \Omega) - I_2(1 \ \Omega) - 4 \ V = 0$
and for a clockwise trip around the right-hand

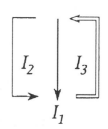

Figure 18.51(b)

mesh, $+4 \ V + I_2(1 \ \Omega) + I_2(5 \ \Omega) + I_3(3 \ \Omega) + I_3(1 \ \Omega) - 12 \ V = 0$
We have three equations in three unknowns.   We reduce by
substitution for I_3 to two equations in two unknowns:

$$\begin{cases} (8 \ \Omega)I_1 - (6 \ \Omega)I_2 - 4 \ V = 0 \\ 4 \ V + (6 \ \Omega)I_2 + (4 \ \Omega)(I_1 + I_2) - 12 \ V = 0 \end{cases}$$

$$\begin{cases} I_2 = \left[(8 \ \Omega)I_1 - 4 \ V\right] / 6 \ \Omega \\ -8V + (4 \ \Omega)I_1 + (10 \ \Omega)I_2 = 0 \end{cases}$$

and to one equation in one unknown:

$$-8 \ V + (4 \ \Omega)I_1 + \frac{10}{6}(8 \ \Omega)I_1 - \frac{10}{6}(4 \ V) = 0$$

$$(17.3 \ \Omega)I_1 - 14.7 \ V = 0$$

$$I_1 = 0.846 \ A \text{ down in 8 } \Omega \ \square$$

Solving by substitution, rather than by determinants, has the
advantage that the answers, after the first, come out much more
easily, just as in having kittens.

18.51 (cont)

Thus $I_2 = [(8\ \Omega)(0.846\ A) - 4\ V]/6\ \Omega = 0.462\ A$ down in middle branch □

$I_3 = 0.846\ A + 0.462\ A = 1.31\ A$ up in 12 V □

It is a good idea to check the math by substituting back into the voltage-rule equations. And if you are new at this, it is worth your time to solve the whole problem again, by taking counterclockwise trips around the loops.

18.61 **PROBLEM:** A 4-MΩ resistor and a 3-μF capacitor are connected in a series with a 12-V power supply.

(a) What is the time constant for the circuit?

(b) Express the current in the circuit and the charge on the capacitor as functions of time.

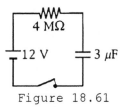

4 MΩ

12 V   3 μF

Figure 18.61

SOLUTION: We suppose the switch is closed at $t = 0$.

(a) $\tau = RC$

$$\tau = (4\ \times\ 10^6\ \Omega)\ (3\ \times\ 10^{-6}\ F)\left(\frac{V}{A\cdot\Omega}\right)\left(\frac{C}{F\cdot V}\right)\left(\frac{A\cdot s}{C}\right)$$

$\tau = 12.0$ s □

(b) $q = \varepsilon C(1 - e^{-t/RC})$

$q = (12\ V)(3\ \mu F)(1 - \exp(-t/12\ s))$

$q = 36.0\ \mu C(1 - \exp(-t/12\ s))$ □

differentiating,

$$I = \frac{dq}{dt} = (36.0\ \mu\ C)\left(0 - \exp\left(-\frac{t}{12\ s}\right)\left(-\frac{1}{12\ s}\right)\right)$$

$I = 3.00\ \mu A\ \exp(-t/12\ s)$ □

18.62 **PROBLEM:** A circuit has been connected as shown in Figure 18.62(a) a "long" time. (a) What is the voltage across the capacitor? (b) If the battery is connected, how long does it take for the capacitor to discharge to 1/10 of its initial voltage?

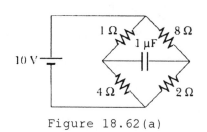

Figure 18.62(a)

SOLUTION: (a) After a long time the capacitor branch will carry negligible current. The current flow is as shown. The voltage rule gives

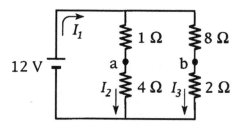

Figure 18.62(b)

$$10 \text{ V} - (1 \ \Omega) I_2 - (4 \ \Omega) I_2 = 0$$

$$I_2 = 2 \text{ A}$$

$$(4 \ \Omega) I_2 = 8 \text{ V}$$

so point a is at 8 V above ground. Similarly,

$$10 \text{ V} - (8 \ \Omega) I_3 - (2 \ \Omega) I_3 = 0 \qquad I_3 = 1 \text{ A} \qquad (2 \ \Omega) I_3 = 2 \text{ V}$$

and point b is at 2 V. The voltage across the capacitor is

$$V_a - V_b = 8 \text{ V} - 2 \text{ V} = 6 \text{ V} \ \square$$

(b) We suppose the battery is pulled out leaving an open circuit. We are left with

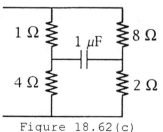

Figure 18.62(c)

18.62 (cont)

equivalent to

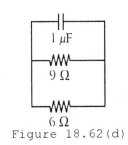

Figure 18.62(d)

in turn equivalent to

$$\cfrac{1}{\cfrac{1}{9}+\cfrac{1}{6}} = 3.6\ \Omega$$

Figure 18.62(e)

So the capacitor sees 3.6 Ω in its discharge.   According to

$$q = Qe^{-t/RC} \qquad qC = QCe^{-t/RC} \qquad V = V_0 e^{-t/RC}$$

$$\frac{1}{10}V_0 = V_0\ e^{-t/3.6\ \Omega 1\ \mu F} \qquad \exp\ (-t/3.6\ \mu s) = 0.1$$

$$(-t/3.6\ \mu s)\ =\ \ln\ 0.1\ =\ -2.30$$

$$+t/3.6\ \mu s\ =\ +2.30$$

$$t\ =\ 2.30\ \text{x}\ 3.6\ \mu s\ =\ 8.29\ \mu s\ \square$$

18.74   PROBLEM: An electric car is designed to run off a bank of 12-V batteries with total energy storage of 2×10^7 J.   (a) If the electric motor draws 8 kW, what is the current delivered to the motor?   (b) If the electric motor draws 8kW as the car moves at a steady speed of 20 m/s, how far will the car travel before it is "out of juice"?

SOLUTION:

(a) $P = VI$ $I = P/V = (8 \times 10^3 \text{ W})/(12 \text{ V}) = 667 \text{ A } \square$

(b)   The car will run for time given by $P = U/t$
    $t = U/P = ((2 \times 10^7 \text{ J})/(8 \times 10^3 \text{ W}))(1 \text{ W} \cdot \text{s/J}) = 2.5 \times 10^3 \text{ s}$
    So it moves distance
    $x = vt = (20 \text{ m/s})(2.5 \times 10^3 \text{ s}) = 50.0 \text{ km } \square$

18.78 **PROBLEM:** An experiment is conducted to measure the electrical resistivity of Nichrome in the form of wires with different lengths and cross-sectional areas. For one set of measurements, a student uses 30-gauge wire, which has a cross-sectional area of 7.3×10^{-8} m^2. The voltage across the wire and the current in the wire are measured with a voltmeter and an ammeter, respectively. For each of the measurements in the following tabel, which were taken on wires of three different lengths, calculate the resistances of the wires and the corresponding values of resistivity. What is the average value of the resistivity, and how does it compare with the value given in Table 18.1?

L(m)	V(V)	I(A)	R(Ω)	r($\Omega \cdot$ m)
0.54	5.22	0.500		
1.028	5.82	0.276		
1.543	5.94	0.187		

SOLUTION: Find each resistance from

$R = V/I$, as in $(5.22$ V$)/(0.5$ A$) = 10.4$ Ω.

Find each resistivity from $R = V/I = \rho \ell /A$

$\rho = VA/I\ell = (5.22$ V$)(7.3 \times 10^{-8}$ m$^2)/(0.5$ A $\times 0.54$ m$)$

$\rho = 1.41 \times 10^{-6}$ $\Omega \cdot$ m

To obtain

L(m)	R(Ω)	r($\Omega \cdot$ m)
0.54	10.4	1.41×10^{-6}
1.028	21.1	1.50×10^{-6}
1.543	31.8	1.50×10^{-6}

18.78 (cont)

Then an average ρ = 1.47 x 10^{-6} $\Omega \cdot m$ $\square$, differing from the tabulated 1.50 x 10^{-6} $\Omega \cdot m$ by 2%. This difference is accounted for by the experimental uncertainty, which we may estimate as (1.47 - 1.41)/1.47 = 4%.

18.79 **PROBLEM:** A general definition of the temperature coefficient of resistivity is $\alpha = \dfrac{1}{\rho}\dfrac{d\rho}{dT}$

where ρ is the resistivity at temperature T.

(a) Assuming that α is constant, show that

$\rho = \rho_0 e^{\alpha(T - T_0)}$

where ρ_0 is the resistivity at temperature T_0. (b) Using the series expansion ($e^x \approx 1 + x;\ x \ll 1$), show that the resistivity is given approximately by the expression $\rho = \rho_0[1 + \alpha(T - T_0)]$ for $\alpha(T - T_0) \ll 1$.

SOLUTION: $a = \dfrac{1}{\rho}\dfrac{d\rho}{dT}$

(a) Separating variables,

$$\int_{\rho_0}^{\rho} \frac{d\rho}{\rho} = \int_{T_0}^{T} \alpha dT$$

$$\ln\left(\frac{\rho}{\rho_0}\right) = \alpha(T - T_0)$$

$$\rho = \rho_0 e^{\alpha(T - T_0)}$$

(b) From the series expansion $e^x \approx 1 + x$, ($x \ll 1$),

$$\rho \approx \rho_0[1 + \alpha(T - T_0)]$$

CHAPTER 19

QUESTIONS

3   **QUESTION:**   A current-carrying conductor experiences no magnetic force when placed in a certain manner in a uniform magnetic field.   Explain.

ANSWER: If the conductor is oriented so that the current is in the same direction as the magnetic field, or in just the opposite direction, then $\sin \theta = 0$ in $\lambda \times \mathbf{B}$, and the force is zero.

8   **QUESTION:**   Will a nail be attracted to either pole of a magnet?   Explain what is happening inside the nail.

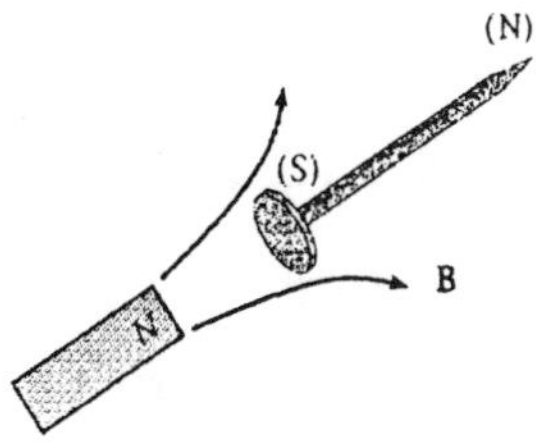

ANSWER: An originally unmagnetized iron nail is attracted to either pole of a magnet.   The magnet's field turns atoms inside the nail to align their magnetic moments with the external field.   The nail becomes an induced magnet.   Since the external field is not uniform, the nail feels a net force of attraction to the magnet.

Figure 19.8

10   **QUESTION:**   A charged particle moves in a circular path because of
     an applied magnetic field. Does the particle gain energy from
     the magnetic field? Explain.

     ANSWER: The particle gains no energy from the magnetic field and
     loses no energy to it, because the q**v** x **B** force is perpendicular
     to the velocity, perpendicular to the incremental displacement,
     and does no work on the particle. The magnetic force never
     speeds up or slows down the motion, but only changes its
     direction.

17   **QUESTION:**   Consider an electron near the magnetic equator. In
     which direction will it tend to be deflected if its velocity is
     directed (a) downward?   (b) northward?   (c) westward?   (d)
     southeastward?

     ANSWER: The magnetic field near the equator is horizontally
     north.
     (a)   If the velocity is down, **v** x **B** is east and q**v** x **B** for the
           negative electron is <u>west</u>.
     (b)   If **v** is north, **v** x **B** is zero and the electron is not
           deflected.
     (c)   If **v** is west, **v** x **B** is west x north in direction, namely
           down, and the electron is deflected <u>up</u> by the q**v** x **B** force.
     (d)   If **v** is southeast, **v** x **B** is in direction southeast x north
           = up and q**v** x **B** deflects the electron <u>down</u>.

Down x North   West x North   SouthEast x North
  is East   is Down   is Up
                    Figure 19.17

24   **QUESTION:** The electron beam in Figure 19.29 is projected to the right. The beam deflects downward in a 0.001-T magnetic field that is produced by a pair of current-carrying coils. (a) What is the direction of the magnetic field? (b) If the diameter of the sphere is 0.1 m, estimate the speed of the electrons in the beam.

ANSWER: The $q\mathbf{v} \times \mathbf{B}$ force on the electrons is down. Since electrons are negative, $\mathbf{v} \times \mathbf{B}$ must be up. With $\mathbf{v}$ to the right, $\mathbf{B}$ must be into the page, away from you. This makes $\mathbf{v} \times \mathbf{B}$ have direction right x away from you = up.

Figure 19.24

(b)   Sketch perpendicular radii to the beam near both ends to locate their intersection as about 10 cm below the starting point. Then $mv^2/r = qvB \sin 90$ gives

$$v = \frac{qBr}{m} \cong \frac{(1.6 \times 10^{-19} \text{ C})(10^{-3} \text{ N}\cdot\text{s/ C}\cdot\text{m})(10^{-1} \text{ m})}{9.11 \times 10^{-31} \text{ kg}}$$

$$v \cong 2 \times 10^7 \text{ m/s}$$

CHAPTER 19

PROBLEMS

19.5 **PROBLEM:** A proton moves perpendicularly to a uniform magnetic field, **B**, with a speed of 10^7 m/s and experiences an acceleration of 2×10^{13} m/s^2 in the +x direction when its velocity is in the +z direction. Determine the magnitude and direction of the field.

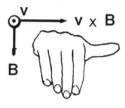

Figure 19.5

SOLUTION:

$F = ma = (1.67 \times 10^{-27} \text{ kg})(2 \times 10^{13} \text{ m/s}^2)$

$F = 3.34 \times 10^{-14} \text{ N} = qvB \sin 90°$

$B = \dfrac{F}{qv} = \dfrac{3.34 \times 10^{-14} \text{ N}}{(1.6 \times 10^{-19} \text{ C})(1 \times 10^7 \text{ m/ s})}$ □

$B = 2.09 \times 10^{-2}$ T

The right hand rule shows that **B** must be in the -y direction to yield a force in the +x direction when v is in the +z direction.

19.7 **PROBLEM:** A cosmic-ray proton in interstellar space has an energy of 10 MeV and executes a circular orbit with a radius equal to that of Mercury's orbit around the Sun (5.8×10^{10} m). What is the galactic magnetic field in that region of space?

SOLUTION: Think of the proton as having accelerated through a potential difference $V = 10^7$ V. Then its energy is

$$E = \frac{1}{2} mv^2 = eV, \text{ so its speed is } v = \sqrt{\frac{2eV}{m}}$$

Now $\Sigma F = ma$ becomes $\dfrac{mv^2}{R} = evB \sin 90°$, so

$$B = \frac{mv}{eR} = \frac{m}{eR} \sqrt{\frac{2eV}{m}} = \frac{1}{R} \sqrt{\frac{2mV}{e}}$$

$$B = \frac{1}{5.8 \times 10^{10} \text{ m}} \sqrt{\frac{2(1.6727 \times 10^{-27} \text{ kg})(10^7 \text{ V})}{1.6 \times 10^{-19} \text{ C}}} = 7.88 \times 10^{-12} \text{ T } \square$$

19.10 **PROBLEM:** The picture tube in a television uses magnetic deflection coils rather than electric deflection plates. Suppose an electron beam is accelerated through a 50-kV potential difference and then passes, for 1 cm, through a uniform magnetic field produced by these coils. The screen is 10 cm from the center of the coils and is 50 cm wide. When the field is turned off, the electron beam hits the center of the screen. What field strength is necessary to deflect the beam to the side of the screen?

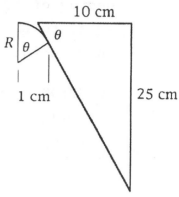

Figure 19.10

SOLUTION: The beam is deflected by the angle

$$\theta = \tan^{-1}\frac{25}{10} = 68.2°$$

The two angles θ shown are equal because their sides are perpendicular, right side to right side and left side to left side. The radius of curvature of the electrons in the field is

$$R = \frac{1.0 \text{ cm}}{\sin 68.2°} = 1.077 \text{ cm}$$

Now $\frac{1}{2}mv^2 = qV$ gives for the speed

$$v = \sqrt{\frac{2qV}{m}} = 1.33 \times 10^8 \text{ m/s, where we choose to ignore the}$$

relativistic correction. At last

$\Sigma F = ma$ becomes $\dfrac{mv^2}{R} = |q|vB \sin 90°$

$$B = \frac{mv}{|q|R} = \frac{(9.11 \times 10^{-31} \text{ kg})(1.33 \times 10^8 \text{ m / s})}{(1.6 \times 10^{-19} \text{ C})(1.077 \times 10^{-2} \text{ m})}$$

$B = 70.1 \text{ mT}$ □

19.14 **PROBLEM:** A wire with a linear mass
density of 0.5 g/cm carries a 2-A
current horizontally to the south.
What are the direction and magnitude
of the minimum magnetic field needed
to lift this wire vertically upward?

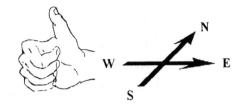

Figure 19.14

SOLUTION: $\dfrac{m}{\ell} = \left(0.5\dfrac{g}{cm}\right)\left(\dfrac{kg}{1000\ g}\right)\left(\dfrac{100\ cm}{m}\right) = 5 \times 10^{-2}$ kg/m

$I = 2$ A in a direction given by right hand rule: eastward

$F = I\ell B \sin\theta$: must counter balance $w = mg$

$mg = I\ell B \sin\theta$

$\dfrac{m}{\ell}g = IB \sin\theta$

$(5 \times 10^{-2})(9.8) = (2)B \sin 90°$

$B = 0.245$ T $\square$

19.19 **PROBLEM:** A strong magnet is
placed under a horizontal
conducting ring of radius r
that carries current I, as
shown in Figure 19.19. If the

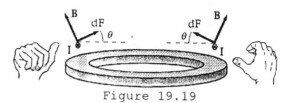

Figure 19.19

magnetic lines of force make an angle of θ with the vertical at
the ring's location, what are the magnitude and direction of
the resultant force on the ring?

SOLUTION: The magnetic force on each bit of ring is $d\mathbf{F} = V \cdot I$
$d\mathbf{s} \times \mathbf{B} = I\, ds\, B$ radially inward and upward, at angle θ above
the radial line. The radially inward components tend to
squeeze the ring but all cancel out as forces. The upward
components $I\, ds\, B \sin\theta$ all add to $I\, 2\pi r B \sin\theta$ up $\square$. The
magnetic moment of the ring is down. This problem is a model
for the force on a dipole in a non-uniform magnetic field, for
the force that one magnet exerts on another magnet.

19.21 **PROBLEM:** A rectangular loop consists of 100 closely wrapped turns and has the dimensions 0.4 m x 0.3 m. The loop is hinged along the y axis, and the plane of the coil makes an angle of 30° with the x axis (Fig. 19.21a). What is the magnitude of the torque exerted on the loop by a uniform magnetic field of 0.8 T, directed along the x axis, when the current in the windings has a value of 1.2 A in the direction shown? What is the expected direction of rotation of the loop?

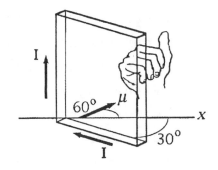

Figure 19.21a

SOLUTION: The magnetic moment of the coil is $\mu = NIA$ perpen-dicular to its plane and making a 60° angle with the x-axis as shown. The torque on the dipole is then $\tau = \mu \times \mathbf{B} = NAIB \sin\theta$ down, having magnitude

$$\tau = N\,B\,AI\,\sin\theta$$

$$\tau = (100)(0.8\ T)(0.4 \times 0.3\ m^2)(1.2\ A)\sin 60°$$

$$\tau = 9.98\ N\cdot m\ \square$$

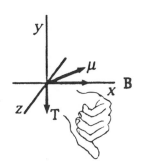

Figure 19.21b

Note θ is the angle between the magnetic moment and the **B** field. Loop will rotate such to align the magnetic moment with the **B** field. Looking down along the y-axis, loop will rotate in the clockwise direction.

19.27 **PROBLEM:** A conduct in the shape of a
square of edge length l = 0.4 m carries
a current of I = 10 A (Fig. 19.27).
Calculate the magnitude and direction
of the magnetic field produced at the
center of the square.

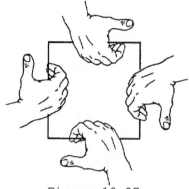

SOLUTION: We use equation 19.14 for
the field created by each side of the

Figure 19.27

square. Each side creates field away from you, into the paper,
in the same direction, so together they produce a field

$$B = \frac{K_m 4 \mu_0 I}{4\pi a}\left(\cos \frac{\pi}{4} - \cos \frac{3\pi}{4}\right)$$

$$= \frac{4 \times 10^{-6}}{0.2}\left(\frac{\sqrt{2}}{2} + \frac{\sqrt{2}}{2}\right) = 2\sqrt{2} \times 10^{-5} \text{ T}$$

B = 28.3 μT away from you into the paper □

19.33 PROBLEM: Two long parallel conductors, separated by a distance of $a = 10$ cm, carry currents in the same direction. If $I_1 = 5$ A and $I_2 = 8$ A, what is the force per unit length exerted on each conductor by the other?

SOLUTION: Take the x-direction to be the direction of the currents and the y-axis to point from 8 A wire toward 5 A wire. At the location of 8 A current, the $I_1 = 5$ A current creates a magnetic field

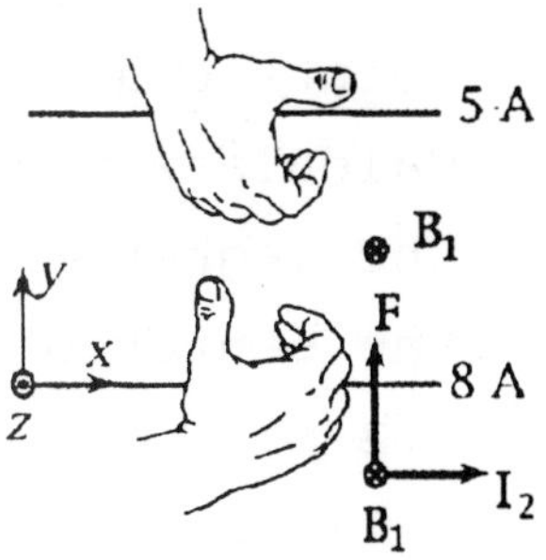

Figure 19.33

$$\mathbf{B}_1 = \frac{\mu_0 I_1}{2\pi a}(-\mathbf{k})$$

Then the force on I_2 is

$$\mathbf{F} = I_2 \ell \times \mathbf{B} = I_2 \ell \frac{\mu_0 I_1}{2\pi a}\mathbf{i} \times (-\mathbf{k})$$

$$\mathbf{F} = \frac{\mu_0 I_1 I_2 \ell}{2\pi a}(\mathbf{j}).$$

This is a force of attraction and the force-per-length is

$$\frac{F}{\ell} = \frac{\mu_0 I_1 I_2}{2\pi a} = 8.00 \times 10^{-5} \text{ N/m} \ \square \text{ on each wire.}$$

19.37 **PROBLEM:**   Four long, parallel conductors carry equal currents,

I = 5 A.   Figure 19.37 is an end view of
the conductors.   The current direction is
into the page at points A and B (indicated
by the crosses) and out of the page at
points C and D (indicated by the dots).
Calculate the magnitude and direction of
the magnetic field at point P, located at
the center of the square of edge length
0.2 m.

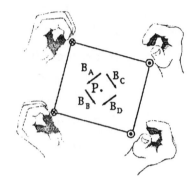

Figure 19.37

SOLUTION: Each wire is distant from P by (0.2 m) cos 45°
= 0.141 m.   Each wire produces a field at P of equal magnitude:

$$B_A = \frac{\mu_0 I}{2\pi a} = \frac{(2 \times 10^{-7} \text{ T} \cdot \text{m})(5 \text{ A})}{A(0.141 \text{ m})} = 7.07 \text{ } \mu T$$

Carrying currents into the page, A produces at P a field of
7.07 µT to the left and down at 135°, while B creates a field
to the right and down, at 45°: carrying currents toward you, C
produces a field downward and to the right, at -45°, while D's
contribution is downward and to the left.   The total field is
then 4 x (7.07 µT) sin 45° = 20.0 µT toward the bottom of the
page.

19.45 **PROBLEM:**   What current is required in the windings of a long
solenoid that has 1000 turns uniformly distributed over a
length of 0.4 m, in order to produce a magnetic field of
magnitude 1.0 x 10^{-4} T at the center of the solenoid?

SOLUTION: $B = \mu_0 \frac{N}{\ell} I$

$$I = \frac{B}{\mu_0 n} = \frac{(10^4 \text{ T} \cdot \text{A})(0.4 \text{ m})}{(4\pi)(10^7 \text{ T} \cdot \text{m})(1000)} = 31.8 \text{ mA} \ \square$$

19.51 PROBLEM: The magnetic moment of the Earth is approximately $8.7 = 10^{22}$ A · m^2. (a) If this were caused by the complete magnetization of a huge iron deposit, how many unpaired electrons would this correspond to? (b) At two unpaired electrons per iron atom, how many kilograms of iron would this correspond to? (The density of iron is 7900 kg/m^3, and there are approximately 8.5 x 10^{28} iron atoms in each cubic meter.)

SOLUTION: The magnetic moment of each unpaired electron is the Bohr magneton,

$$9.27 \times 10^{-24} \ J/T = \frac{9.27 \times 10^{-24} \ J}{T} \left(\frac{N \cdot m}{J}\right)\left(\frac{T \cdot C \cdot m}{N \cdot S}\right)\left(\frac{A \cdot s}{C}\right)$$

$$u_B = 9.27 \times 10^{-24} \ A \cdot m^2$$

(a) Number of unpaired electrons $= \dfrac{8.7 \times 10^{22} \ A \cdot m^2}{9.27 \times 10^{-24} \ A \cdot m^2}$

$N = 9.39 \times 10^{45}$ □

Each iron atom has two unpaired electrons, so the number of iron atoms required is ½(9.39 x 10^{45}).

(b) Mass $= \dfrac{4.69 \times 10^{45} \left(7900 \ kg/ \ m^3\right)}{8.5 \times 10^{28} \ atoms/ \ m^3} = 4.36 \times 10^{20}$ kg □

19.53 **PROBLEM:** Measurements of the magnetic field of a large tornado were made at the Geophysical Observatory in Tulsa, Oklahoma, in 1962. If the tornado's field was **B** = 1.5 x 10^{-8} T, pointing north, when the tornado was 9 km east of the observatory, what current was carried up or down the funnel of the tornado?

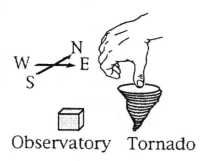

Observatory Tornado

Figure 19.53

SOLUTION: Consider the funnel as containing a long straight vertical current, to produce field $B = \mu_0 I/2\pi r$, so

$$I = \frac{2\pi r B}{\mu_0} = \frac{2\pi(9000)\left(1.5 \times 10^{-8}\right)}{4\pi \times 10^{-7}} = 675 \text{ A } \square$$

(Conventional) current is downward *or* negative charge flows upward.

19.61 **PROBLEM:** A singly charged heavy ion is observed to complete five revolutions in a uniform magnetic field of magnitude 5 x 10^{-2} T in 1.50 ms. Calculate the (approximate) mass of the ion in kilograms.

SOLUTION: The magnetic force is constant in magnitude, pushing the particle into a path of constant curvature, a circle, and constituting the centripetal force:

$$mv^2/r = |q|vB \sin 90°$$

Then $\dfrac{v}{r} = \omega = \dfrac{qB}{m}$ or $m = \dfrac{qB}{\omega} = \dfrac{qB}{2\pi f}$

$$m = \frac{(1.6 \times 10^{-19} \text{ C})(5 \times 10^{-2} \text{ T})}{(2\pi)(5 \text{ rev}/ 1.5 \times 10^{-3} \text{ s})} = 3.82 \times 10^{-25} \text{ kg } \square$$

19.62 **PROBLEM:** Sodium melts at 210°F. Liquid sodium, and excellent thermal conductor, is used in some nuclear reactors to remove thermal energy from the reactor core. The liquid sodium can be moved through pipes by pumps that exploit the force on a moving charge in a magnetic field. The principle is as follows: imagine the liquid metal to be in a pipe having a retangular cross-section of width w and height h. A uniform magnetic field perpendicular to the pipe affects a section of length L(Fig. 19.62). An electric current directed perpendicular to the pipe and to the magnetic field produces a current density of **J**. (a) Explain why this arrangement produces a force on the liquid that is directed along the length of the pipe. (b) Show that the section of liquid in the magnetic field experiences a pressure increase equal to JLB.

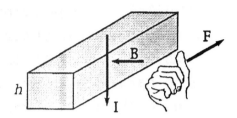

Figure 19.62

SOLUTION:

(a) The electric current carried by the material experiences a force $I\mathbf{h} \times \mathbf{B}$ in the direction of L.

(b) The sodium, consisting of ions and electrons, flows along the pipe transporting no net charge. But inside the section of length L electrons drift upward to constitute downward electric current $\mathbf{J} \times$ (area) $= \mathbf{J}lw$. The current a magnetic force $|I\mathbf{h} \times \mathbf{B}| = JLwhB \sin 90°$. This force along the pipe axis will make the fluid move, exerting pressure

$$\frac{F}{\text{area}} = \frac{JLwhB}{hw} = JLB$$

In the picture, the fluid moves away from you into the page.

19.67 **PROBLEM:** Two circular loops are
parallel, co-axial, and almost in
contact, 1 mm apart (Fig. 19.67).
Each loop is 10 cm in radius. The top
loop carries a current of 140 A clockwise. The bottom loop carries 140 A
counter-clockwise. (a) Calculate the
magnetic force that the bottom loop

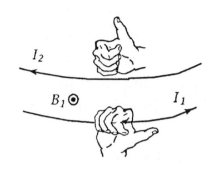

Figure 19.67

exerts on the top loop. (b) The upper loop has a mass of
0.021 kg. Calculate its acceleration, assuming that the only
forces acting on it are the force in part (a) and its weight.

SOLUTION: Note 1 mm is small compared to 10 cm, so the lower
wire creates a field $B_1 = \mu_0 I_1 / 2\pi a$ as a long straight wire
does. At the near point of the loop, the lower wire creates
field toward you in the space above it. The upper wire then
experiences a force $I_2 \ell \times \mathbf{B}_1$ of repulsion, amounting to

(a) $F = \dfrac{\mu_0 I^2 L}{2\pi a}$ (Equation 19.20)

$$F = \frac{\left(4\pi \ \times \ 10^{-7}\right)\left(140^2\right)(2\pi)\,(0.1)}{2\pi\left(10^{-3}\right)} = 2.46 \text{ N} \ \square$$

(b) $a_{loop} = \dfrac{2.46 \text{ N} - m_{loop}\,g}{m_{loop}} = 107.3 \text{ m/s}^2 \uparrow \ \square$

CHAPTER 20

QUESTIONS

8   **QUESTION:** How is electrical energy produced in dams (that is, how is the energy of motion of the water converted to ac electricity)?

ANSWER: The falling water pushes on the vanes of a turbine functioning as a waterwheel, to turn it and with it the rotating coil of a generator. In the coil the emf induced is the source of the voltage driving current throughout the electric power grid. When current flows in the generator coil a backward magnetic force acts on each wire. It is against this force that the water must do work.

12   **QUESTION:** When the switch in the circuit shown in Figure 20.12 is closed, a current is set up in the coil and the metal ring springs upward (see Figure 20.12). Explain this behavior.

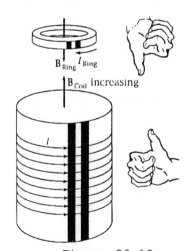

Figure 20.12

ANSWER: Closing the switch makes rapidly increasing current flow counterclockwise in the coil, so the coil becomes an electromagnet with rapidly increasing upwardfield. As a magnetic dipole its N pole is at the top and its S pole at the bottom. The ring encircles increasing upward flux and so has induced in it an emf to produce clockwise current to produce a downward magnetic field. The ring becomes a dipole with N pole at the bottom and S pole at the top. The repulsion of the adjacent N poles is responsible for the upward force in the ring.

17   **QUESTION:**   Find the direction of
the current through the resistor in
Figure 20.17, (a) at the instant
the switch is closed.   (b) after
the switch has been closed for
several minutes, and (c) at the
instant the switch is opened.

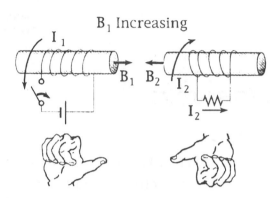

Figure 20.17a

ANSWER: (a)   The battery makes
counterclockwise current I_1 in
the primary coil, so its magnetic
field $\mathbf{B}_1$ is to the right and
increasing just after the switch
is closed.   The secondary coil
will oppose the change with a
leftward field $\mathbf{B}_2$, which comes
from an induced clockwise current
I_2 that goes to the <u>right</u> in the
resistor.

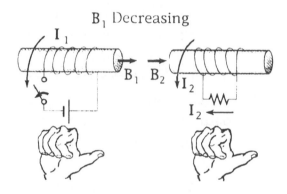

Figure 20.17b

(b)   At steady state the primary magnetic field is unchanging, so
<u>no</u> emf is induced in the secondary.

(c)   The primary's field is to the right and decreasing as the
switch is opened.   The secondary opposes the decrease by making
its field to the right, carrying counterclockwise current to the
left in the resistor.

CHAPTER 20

PROBLEMS

20.3   **PROBLEM:**   The plane of a retangular coil having dimensions of 5 cm x 8 cm is perpendicular to the direction of a magnetic field, **B**.   If the coil has 75 turns and a total resistance of 8 Ω, at what rate must the magnitude of **B** change in order to induce a current of 0.1 A in the windings of the coil?

SOLUTION: The emf generated in the coil is

$$\mathcal{E} = -N\frac{d}{dt}(BA)\cos\theta = -NA1\frac{dB}{dt} , \text{ to produce current}$$

$$I = \frac{|\mathcal{E}|}{R} = \frac{NA}{R}\frac{dB}{dt}$$

$$\frac{dB}{dt} = \frac{IR}{NA} = \frac{(0.1 \text{ A})(8 \text{ }\Omega)}{75(40 \times 10^{-4} \text{ m}^2)}\left(\frac{V}{A\cdot\Omega}\right)\left(\frac{N\cdot m}{V\cdot C}\right)\left(\frac{T\cdot cm}{N\cdot s}\right) = 2.67 \text{ T/s } \square$$

20.7   **PROBLEM:**   A long, straight wire carries a current that varies with time as $I = I_0\sin(\omega t + \delta)$ and lies in the plane of a rectangular loop of N turns of wire, as shown in Figure 20.7.   The quantities I_0, ω, and δ are all constants.   Determine the emf induced in the loop by the magnetic field due to the current in the straight wire. Assume $I_0 = 50$ A, $\omega = 200\pi s^{-1}$, $N = 100$, $a = b = 5$ cm, and $\ell = 20$ cm.

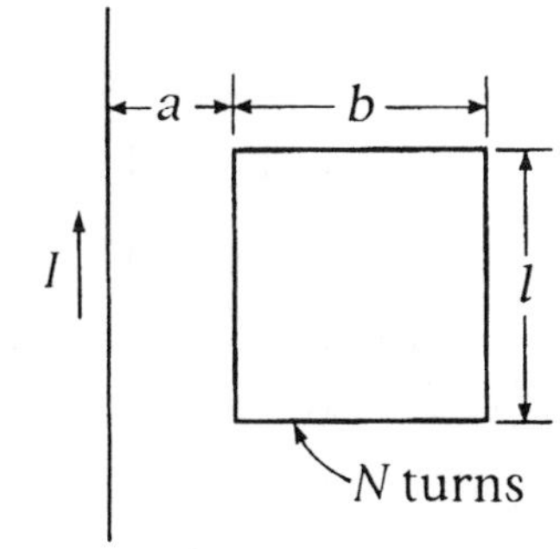

Figure 20.7

20.7 (cont.)

SOLUTION: The loop is the boundary of the plane surface S. The magnetic field produced by the current in the straight wire is perpendicular to S at all points on the surface. The magnitude of the field is

$B = \dfrac{\mu_0 I}{2\pi r}$. Thus the flux linkage is

$\Phi_m = N\Phi_m$

$= \dfrac{\mu_0 NI\ell}{2\pi} \int_{a}^{a+b} \dfrac{dr}{r} = \dfrac{\mu_0 NI_0 \ell}{2\pi} \ln\left(\dfrac{a+b}{a}\right) \sin(\omega t + \delta)$

Finally, the induced EMF is in absolute value

$|\varepsilon| = \dfrac{d\,\Phi_m}{dt} = \dfrac{\mu_0 NI_0 \ell\omega}{2\pi} \ln\left(\dfrac{a+b}{a}\right) \cos(\omega t + \delta)$

$= (2 \times 10^{-7}\,\text{T}\cdot\text{m/ A})(100)(50\,\text{A})(0.20\,\text{m})\left(200\pi\,\text{s}^{-1}\right) \ln\left(\dfrac{5+5}{5}\right) \cos(\omega t + \delta)$

$|\varepsilon| = (87.1\,\text{mV}) \cos(200pt + d)$ □

The term $\sin(\omega t + \delta)$ in the expression for the current in the straight wire does not change appreciably when ωt changes by 0.1 rad or less. Thus, the current does not change appreciably during a time interval $t < 0.1/(200\pi\text{s}^{-1}) = 1.6 \times 10^{-4}$ s. We define a critical length, $ct = (3 \times 10^{8}\,\text{m/s})(1.6 \times 10^{-4}\,\text{s}) = 4.8 \times 10^{4}$ m, equal to the distance to which field changes could be propagated during an interval of 1.6×10^{-4} s. This length is so much larger than any dimension of the loop or its distance from the wire that, although we consider the straight wire to be infinitely long, we can also safely ignore the field propagation effects in the vicinity of the loop. Moreover, the phase angle can be considered to be constant along the wire in the vicinity of the loop.

20.7   (cont.)

If the frequency ω were much larger, say, $200\pi \times 10^5$ s^{-1}, the corresponding critical length would be only 48 cm. In this situation propagation effects would be important and the above expression of $\mathcal{E}$ would require modification. As a "rule of thumb" we can consider field propagation effects for circuits of laboratory size to be negligible for frequencies, $f = \omega/2\pi$, that are less about 10^6 Hz.

20.19   **PROBLEM:** A conducting rectangular loop of mass M, resistance R, and dimensions w wide by ℓ long falls from rest into a magnetic field, **B**, as shown in Figure 20.19. The loop accelerates until it reaches terminalspeed, v_t. (a) Show that $v_t = \dfrac{MgR}{b^2w^2}$. (b) Why is v_t proportional to R? (c) Why is it inversely proportional to B^2?

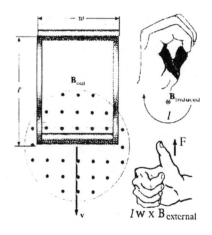

Figure 20.19

20.19 (cont)

    SOLUTION: Let y represent the vertical dimension of the lower part of the loop where it is inside the strong magnetic field. As the loop falls y increases; the loop encloses increasing flux toward you and has induced in it an emf to produce a current to make its own magnetic field away from you. This current is to the left in the bottom side of the loop, and feels an upward force in the external field. Symbolically, the flux is

$\Phi = BA \cos \theta = Bwy \cos 0$. The emf is

$$|\varepsilon| = -N \frac{d}{dt} Bwy = -1Bw \frac{dy}{dt} = -Bwv.$$

The magnitude of the current is

$I = |\varepsilon|/R = Bwv/R$, and the force is

$$\mathbf{F} = I\mathbf{w} \times \mathbf{B} = (Bwv/R)wB \sin 90° \text{ up}$$
$$= B^2 w^2 v/R \text{ up}$$

At terminal speed the loop is in equilibrium:

$$|\varepsilon| Fy = 0 \qquad +B^2 w^2 V_t/R - Mg = 0 \qquad \therefore V_t = MgR/B^2 w^2$$

(b)   The EMF is directly proportional to v, but the current is inversely proportional to R. A large R means a small current at a given speed, so the loop must travel faster to get F_{mag} = weight.

(c)   At a given speed, the current is directly proportional to the magnetic field. But the force is proportional to the product of the current and the field. For a small B, the speed must increase to compensate for both the small B and also the current, so $v \propto B^{-2}$.

20.23 **PROBLEM:** A square coil (20 cm x 20 cm) that consists of 100 turns of wire rotates about a vertical axis at 1500 rpm, as indicated in Figure 20.23. The horizontal component of the Earth's magnetic field at the location of the loop is 2×10^{-5} T. Calculate the maximum emf induced in the coil by the Earths field.

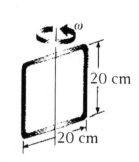

SOLUTION: Let θ represent the angle through

Figure 20.23

which the coil turns, starting from $\theta = 0$ at an instant when the horizontal component of the earth's field is perpendicular to the area. Then

$$\mathcal{E} = - N \frac{d}{dt} BA \cos \theta = -NBA \frac{d}{dt} \cos \omega t$$

$$\mathcal{E} = +NBA \omega \sin \omega t$$

Here $\sin \omega t$ oscillates between +1 and -1, so the spinning coil generates an alternating voltage with amplitude

$$\mathcal{E}_{max} = NBA \omega = NBA \ 1\pi f$$

$$\mathcal{E}_{max} = 100 (2 \times 10^{-5} \ T)(0.2 \ m)^2 2\pi (1500) \frac{1}{min} \left(\frac{1 \ min}{60 \ s}\right) = 12.6 \ mV \ \square$$

20.29   **PROBLEM:**   An aluminum ring having a radius
of 5 cm and resistance 3×10^{-4} Ω is placed
on top of a long air-core solenoid with
1000 turns per meter and radius 3 cm, as
shown in Figure 20.29.   At the location of
the ring, the magnetic field due to the
current in the solenoid is one-half that at
the center of the solenoid.   If the current
in the solenoid is *increasing* at a rate of
270 A/s, (a) what is the induced current in the ring?   (b) At the
center of the ring what is the magnetic field produced by the
induced current in the ring?   (c) What is the direction of the
field in (b)?

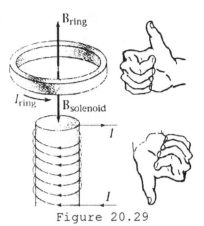

Figure 20.29

SOLUTION:

$$\mathcal{E} = - N \frac{d}{dt} BA \cos \theta = -1 \frac{d}{dt} 0.5 \, \mu_0 nIA \cos 0$$

$$\mathcal{E} = - 0.5 \, \mu_0 nA \frac{dI}{dt}$$

Note that A must be interpreted as the area of the solenoid,
where the field is strong:

$$\mathcal{E} = -0.5 (4\pi \times 10^{-7} \text{ T} \cdot \text{m/A}) (10^3/\text{m}) \pi (0.03 \text{ m})^2 270 \text{ A/s}$$

$$(N \cdot s/C \cdot m \cdot T) (V \cdot C/N \cdot m)$$

$$\mathcal{E} = -4.8 \times 10^{-4} \text{ V}$$

(a) $I_{ring} = \dfrac{|\mathcal{E}|}{R} = \dfrac{0.00048}{0.0003} = 1.60 \text{ A} \; \square$

(b) $B_{ring} = \dfrac{\mu_0 I}{2R} = 2.01 \times 10^{-5} \text{ T} \; \square$

(c)   The coil's field points downward, and is increasing, so
B_{ring} points upward $\square$.

20.31 **PROBLEM:** A proton moves through a uniform electric field given by **E** = 50**j** V/m and a uniform magnetic field, **B** = (0.2**i** + 0.3**j** + 0.4**k**) T. Determine the acceleration of the proton when it has a velocity given **v** = 200**i** m/s.

SOLUTION: **F** = m**a** = q**E** + q**v** × **B**

$$\mathbf{a} = \frac{e}{m}[E + v \times B] \quad \text{where}$$

$$\mathbf{v} \times \mathbf{B} = \begin{vmatrix} \mathbf{i} & \mathbf{j} & \mathbf{k} \\ 200 & 0 & 0 \\ 0.2 & 0.3 & 0.4 \end{vmatrix} = -200(0.4)\mathbf{j} + 200(0.3)\mathbf{k}$$

$$\mathbf{a} = \frac{1.6 \times 10^{-19}}{1.67 \times 10^{-27}}[50\,\mathbf{j} - 80\,\mathbf{j} + 60\,\mathbf{k}] = 9.58 \times 10^{7}[-30\mathbf{j} + 60\mathbf{k}]$$

$$\mathbf{a} = 2.87 \times 10^{9}(-\mathbf{j} + 2\mathbf{k}]\text{m/s}^2 = (-2.87 \times 10^{9}\,\mathbf{j} + 5.75 \times 10^{9}\,\mathbf{k})\ \text{m/s}^2 \ \square$$

20.39 **PROBLEM:** A solenoidal inductor contains 420 turns, is 16 cm in length, and has a cross-sectional area of 3 cm^2. What uniform rate of decrease of current through the inductor will produce an induced emf of 175 μV?

SOLUTION: $L = \dfrac{\mu_0 N^2 A}{\ell} = \dfrac{\mu_0 (420)^2 (3 \times 10^{-4})}{0.16} = 4.16 \times 10^{-4}$ H

$$\varepsilon = -L\frac{dI}{dt} \rightarrow \frac{dI}{dt} = \frac{-\varepsilon}{L} = \frac{-175 \times 10^{-6}\ \text{V}}{4.16 \times 10^{-4}\ \text{H}}$$

$$\varepsilon = -0.421\ \text{A/s}$$

20.43 **PROBLEM:** A 12-V battery is about to be connected to a series circuit containing a 10-W resistor and a 2-H inductor. (a) How long will it take the current to reach 50% of its final value? (b) How long will it take to reach 90% of its final value?

SOLUTION: The time constant is $\tau = \dfrac{L}{R} = 0.2s$

(a) In $I = (\mathcal{E}/R)(1 - e^{-t/\tau})$ the final value which the current approaches is $I = (\mathcal{E}/R)(1 - e^{-\infty}) = \mathcal{E}/R$. We have

$0.50\ \mathcal{E}/R = (\mathcal{E}/R)(1 - e^{t/0.2})$

$0.50 = 1 - \exp(-t/0.2)$

$\exp(-t/0.2) = 0.50$

$\exp(t/0.2) = 2$

$t/0.2 = \ln 2$

$\therefore\ t = 0.139\ \text{s}\ \square$

(b) $0.90 = 1 - e^{-t/0.2} \rightarrow t = \tau \ln 10 = 0.461\ \text{s}\ \square$

20.51 **PROBLEM:** A 10-V battery, a 5-Ω resistor, and a 10-H inductor are connected in series. After the current in the circuit has reached its maximum value, calculate (a) the power supplied to the circuit by the battery, (b) the power dissipated in the resistor, (c) the power dissipated in the inductor, and (d) the energy stored in the magnetic field of the inductor.

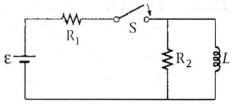

Figure 20.51

SOLUTION: From Equation 20.19, $I = \dfrac{\mathcal{E}}{R}\left(1 - e^{-Rt/L}\right)$

(a) The maximum current is reached after a long time t and is

$I = \dfrac{\mathcal{E}}{R} = 2$ A. At that time, the inductor is fully energized and the battery output is

$P = IV = (2\ \text{A})(10\ \text{V}) = 20\ \text{W}$ □

(b) $P_{lost} = I^2R = (2\ \text{A})^2\,(5\ \Omega) = 20\ \text{w}$ □

(c) $P_{inductor} = IV_{drop} = 0$ □

(d) $E_{stored} = \dfrac{LI^2}{2} = \dfrac{(10\ \text{H})(2\ \text{A})^2}{2} = 20\ \text{J}$ □

20.53 **PROBLEM:** The magnetic field inside a supersonducting solenoid is 4.5 T. The solenoid has an inner diameter of 6.2 cm and a length of 26 cm. (a) Determine the magnetic energy density in the field. (b) Determine the magnetic energy stored in the magnetic field within the solenoid.

SOLUTION:

(a) The magnetic energy density is given by Equation 20.27,
$u_M = B^2/2\mu_0 = (4.5\ \text{T})^2/2(1.26 \times 10^{-6}\ \text{T}\cdot\text{m/A})$
$u_M = 8.06 \times 10^6\ \text{J/m}^3$ □

(b) The magnetic energy stored in the fields equals u_M times the volume of the solenoid (the volume in which B is non-zero).
$U_M = u_M V = (8.06 \times 10^6\ \text{J/m}^3)\{(0.26\ \text{m})\pi(0.031\ \text{m})^2\} = 6.32\ \text{kJ}$ □

20.61   **PROBLEM:**   A long solenoid with 1000 turns/m and a radius 2 cm carries an oscillating current given by the expression $I = (5\ A)\ \sin\ (100\ \pi t)$. What is the electric field induced at a radius $r = 1$ cm from the axis of the solenoid? What is the direc-tion of this electric field when the current is *increasing* counter-clockwise in the coil?

SOLUTION: In $\oint \mathbf{E} \cdot d\mathbf{s} = -\dfrac{d}{dt} \oint \mathbf{B} \cdot d\mathbf{a}$,   consider a circle of radius 1 cm containing the point in question. The induced electric field is uniform in strength around this circle and everywhere tangent to it. The magnetic field is uniform within it:

$$E \cos 0 \oint ds = -\frac{d}{dt} B \oint da$$

$$E\, 2\pi r = -\frac{d}{dt}\ \mu_0 n I \pi\ r^2$$

$$E = -\left(\mu_0 n r\ /\ 2\right)\frac{d}{dt}\ 5\ A \sin(100\pi t)$$

$$E = -(\mu_0 n r/2)\, 5\ A(100\pi/s) \cos(100\pi t)$$

$$E = -(4\pi \times 10^{-7}\ N/A^2)\,(10^3/m)\,(10^{-2}m/2)\, 5\ A(100\pi/s)$$

$$(As/C)\cos(100\pi t)$$

$E = -(9.87\ mN/C)\cos(100\pi t)$ □

Carrying counterclockwise current, the solenoid creates upward magnetic field in the picture. As the magnetic field increases, the electric field opposes the change by being clockwise o, so that if there <u>were</u> a wire of radius one centimeter it would carry clockwise current to make a downward magnetic field of its own.

20.65 PROBLEM: A long solenoid has n turns per meter and carries a current of $I = I_0(1 - e^{-at})$, with $I_0 = 30$ A and $\alpha = 1.6$ s^{-1}. Inside the solenoid and coaxial with it is a loop that has a radius of $R = 6$ cm and consists of a total of N turns of fine wire. What emf is induced in the loop by the changing current? Take $n = 400$ turns/m and $N = 250$ turns. (Assume that the loop is at the center of the solenoid, where the magnetic field is uniform and perpendicular to the plane of the loop.)

SOLUTION: The solenoid creates magnetic field $B = \mu_0 nI$ $= \mu_0 nI_0(1 - e^{-at})$. The flux through one turn of the loop is

$$\Phi_m = \int BdA \cos 0$$

$$\Phi_m = \mu_0 nI_0(1 - e^{-at})\int dA$$

$$\Phi_m = \mu_0 nI_0(1 - e^{-at})\pi R^2. \quad \text{The emf generated is}$$

$$\mathcal{E} = - N\frac{d\,\Phi_m}{dt} = - N\,\mu_0 nI_0\pi R^2\alpha e^{-at}$$

$$\mathcal{E} = -(250)(4\pi \times 10^{-7}\ \text{N/A}^2)(400\ \text{m}^{-1})(30\ \text{A})$$

$$(\pi(0.06\ \text{m})^2)(1.6\ \text{s}^{-1})e^{-1.6t}$$

$$\mathcal{E} = -68.2\ e^{-1.6t}\ \text{mV}\ \square$$

20.71   **PROBLEM:**   The magnetic flux threading a metal ring varies with
time t according to

$$\Phi_m = 3(at^3 - bt^2) \text{ T} \cdot \text{m}^2 \qquad a = 2 \text{ s}^{-3} \qquad b = 6 \text{ s}^{-2}$$

The resistance of the ring is 3 Ω.   Determine the *maximum*
current induced in the ring during the interval from $T = 0$ to
$t = 2$ s.

SOLUTION: $\Phi_m = (6t^3 - 18t^2) \text{T} \cdot \text{m}^2$

$$\mathcal{E} = -\frac{d \Phi_m}{dt} = -18t^2 + 36t$$

Maximum $\mathcal{E}$ occurs when $\dfrac{d \mathcal{E}}{dt} = -36t + 36 = 0$, which gives $t = 1$ s.

$\therefore$ Maximum current (at $t = 1$ s) is

$$I = \frac{\mathcal{E}}{R} = \frac{(-18 + 36) \text{ V}}{3 \ \Omega} = 6.0 \text{ A} \ \square$$

20.75 **PROBLEM:** At $t = 0$, the switch in Figure 20.75 is closed. By using Kirchhoff's laws for the instantaneous currents and voltages in this two-loop circuit, show that the current through the inductor is $I(t) = \dfrac{\mathcal{E}}{R_1}\left[1 - e^{(-R'/L)t}\right]$ where $R' = R_1R_2/(R_1 + R_2)$.

SOLUTION: Call I the downward current through the inductor and I_2 the downward current through R_2. Then $I + I_2$ is the current in R_1.

Left hand loop: $\mathcal{E} - (I + I_2)R_1 - I_2R_2 = 0$

Outside loop: $\mathcal{E} - (I + I_2)R_1 - L\dfrac{dI}{dt} = 0$

eliminate I_2, obtaining

$$I\underbrace{\left(\frac{R_1R_2}{R_1 + R_2}\right)}_{R^1} + L\frac{dI}{dt} = \underbrace{\left(\frac{R_2}{R_1 + R_2}\right)\mathcal{E}}_{\mathcal{E}^1}$$

$$\therefore \mathcal{E}' - IR' - L\frac{dI}{dt} = 0$$

This is of the same form as Equation 20.18, so the reasoning on page 589 shows that the solution is the same form as Equation

20.19: $I = \dfrac{\mathcal{E}'}{R'}\left(1 - e^{-R't/L}\right)$, with

$$\frac{\mathcal{E}'}{R'} = \frac{\mathcal{E}R_2 / (R_1 + R_2)}{R_1R_2 / (R_1 + R_2)} = \frac{\mathcal{E}}{R'}: \quad I(t) = \frac{\mathcal{E}}{R'}\left(1 - e^{R't/L}\right) \quad \square$$

CHAPTER 21

QUESTIONS

1 **Question:** Does the acceleration of a simple harmonic oscillator
remain constant during its motion? Is the acceleration ever
zero? Explain.

Answer: In SHM the acceleration is not constant. It is zero
whenever the object passes through equilibrium, to the right
whenever the object is on the left of equilibrium, and in general
proportional to the displacement but oppositely directed. So be
sure you think of the equations describing motion with constant
acceleration, $x = x_0 + v_0 t + \frac{1}{2} a t^2$, $v = v_0 + at$,
$v^2 = v_0^2 + 2a(x - x_0)$ as replaced by the equation describing simple
harmonic motion:

$$x = A \cos (\omega t + \phi), \quad v = -\omega A \sin (\omega t + \phi), \quad v = \pm \omega \sqrt{A^2 - x^2}$$

9 **Question:** A pendulum bob is made from a ball filled with water.
 What happens to the frequency of bivration of this pendulum if
the ball has a hole that allows water to slowly leak out?

Answer: Nothing happens to the period, or to the amplitude.
They are independent of the mass of the pendulum. The energy
decreases proportionately with the mass.

16 **Question:** (a) Find the ratio of the period of a pendulum on Earth to the period of an identical pendulum on the Moon, where free-fall acceleration is one-sixth that on Earth. (b) If the period of the pendulum is 2.5 s on Earth, what will be its period on the Moon?

Answer: To compare $T_e = 2\pi\sqrt{L / g_e}$ and $T_m = 2\pi\sqrt{L / g_m}$, divide the equations:

$T_e/T_m = \sqrt{g_m / g_e} = \sqrt{1 / 6} = 0.408$.

(b) $T_m = T_e/0.408 = (2.5 \text{ s})/0.408 = 6.12 \text{ s}$

19 **Question:** The amplitude of a system moving with simple harmonic motion is doubled. Determine the changes in (a) the total energy, (b) the maximum velocity, (c) the maximum acceleration, and (d) the period.

Answer:
(a) Energy is proportional to amplitude squared, so it gets four times larger.
(b) Maximum velocity is proportional to the square root of energy, so it doubles. ($V_{max} = \omega A$)
(c) Acceleration is proportional to displacement, so maximum acceleration doubles. ($a_{max} = \omega^2 A$)
(d) In SHM, as opposed to many other repeating motions, period is independent of amplitude. It is unchanged.

CHAPTER 21

PROBLEMS

21.7   **Problem:** A particle moving along the x axis with simple harmonic motion starts from the origin at $t = 0$ and moves toward
the right.   The amplitude of its motion is 2 cm, and the
frequency is 1.5 Hz.   (a) Show that the displacement of the
particle is given by $x = (2$ cm$) \sin (3\pi t)$.   Determine (b) the
maximum speed and the earlies time $(t > 0)$ at which the particle has this speed, (c) the maximum acceleration and the
earliest time $(t > 0)$ at which the particle has this acceleration, and (d) the total *distance* traveled between $t = 0$ and
$t = 1$ s.

Solution:

(a)   At $t = 0$, $x = 0$ and v is positive (to the right).   The
sine function is zero and the cosine is positive at $\theta = 0$,
so this situation corresponds to $x = A \sin \omega t$ and $v = v_0$
$\cos \omega t$.   Since $f = 1.5$ Hz, $\omega = 2\pi f = 3p$; also, $A = 2$ cm,
so that $x = 2 \sin 3\pi t$ cm ☐.   This is equivalent to writing
$x = A \cos (\omega t + \phi)$ with $A = 2$ cm, $\omega = 3\pi$/s and $\phi = -90° =$
$-\pi/2$.   Note $T = 1/\phi = (2/3)$s.

(b)   The velocity is $v = dx/dt = 2(3\pi) \cos (3\pi t)$.   The maximum
speed is $v_{max} = v_0 = A\omega = 2(3\pi) = 6\pi$ cm/s ☐.   The particle
has this speed at $t = 0$, when $\cos (3\pi t) = +1$, and next at
$t = \dfrac{T}{2} = \dfrac{1}{3}$ s ☐, when $\cos\left(3\pi \dfrac{1}{3}\right) = -1$

21.7 (cont)

(c) Again, $a = dv/dt = -2(3\pi)^2 \sin(3\pi t)$. Its maximum value is $a_{max} = A\omega^2 = 2(3\theta)^2 = 18\pi^2$ cm/s^2 □. The acceleration has this positive value for the first time at $t = \dfrac{3T}{4} = 0.500$ s □, when $a = -2(3\pi)^2 \sin(3\pi/2) = +18\pi^2$

(d) Since $T = \dfrac{2}{3}$ s and $A = 2$ cm, the particle will travel 8 cm in this time. Hence, in $1s \left(\dfrac{3T}{2}\right)$ the particle will travel 8 cm + 4 cm = 12.0 cm □

21.9 **Problem:** A 3-kg mass is attached to a spring and pulled out horizontally to a maximum displacement from equilibrium of 0.5 m. What spring constant must the spring have if the mass is to achieve an acceleration equal to the free-fall acceleration g?

Solution: From Newton's second law, $|\Sigma F| = |ma| = |-kx|$

$a = \dfrac{kx}{m}$, or 9.8 m/s$^2 = k \dfrac{0.5 \text{ m}}{3 \text{ kg}}$

From which, $k = 58.8$ N/m □

21.19 **Problem:** A 50-g mass, connected to a light spring of force
constant 35 N/m, oscillates on a horizontal track with an
amplitude of 4 cm. Friction is negligible. Find (a) the total
energy of the oscillating system and (b) the speed of the mass
when the displacement is 1 cm. When the displacement is 3 cm,
find (c) the kinetic energy and (d) the potential energy.

Solution: (a) $E = \frac{1}{2}kA^2 = \frac{1}{2}(35 \text{ N/m})(0.04 \text{ m})^2 = 0.028 \text{ J}$ □

(b) We express conservation of energy as
$\frac{1}{2}mv^2 + \frac{1}{2}kx^2 = \frac{1}{2}kA^2$. Solving for speed gives

$$|v| = \omega\sqrt{A^2 - x^2} = \sqrt{\frac{k}{m}} \sqrt{A^2 - x^2}$$

$$|v| = \sqrt{\frac{35}{0.05}} \sqrt{(0.04)^2 - (0.01)^2} = 1.02 \text{ m/s} \quad \square$$

(c) $\frac{1}{2}mv^2 = \frac{1}{2}kA^2 - \frac{1}{2}kx^2$

$K = \frac{1}{2}(35)[(0.04)^2 - (0.01)^2] = 12.2 \text{ mJ}$ □

(d) $V = \frac{1}{2}kx^2 = E - \frac{1}{2}mv^2 = 15.8 \text{ mJ}$ □

21.31 **Problem:** When the simple pendulum illustrated
in Figure 21.31 makes an angle of θ with the
vertical, its speed is v. (a) Calculate the
total mechanical energy of the pendulum as a
function of v and θ. (b) Show that when θ is
small, the potential energy can be expressed as
$\frac{1}{2}mgL\theta^2 = \frac{1}{2}m\omega^2 s^2$. (*Hint:* In part (b), use the
small angle approximation $\cos\theta \approx 1 - \theta^2/2$.)

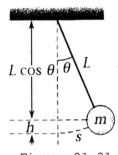

Figure 21.31

Solution: The pendulum bob possesses kinetic energy and
gravitational potential energy, adding to $E = \frac{1}{2}mv^2 + mgh$.
When the pendulum makes an angle θ with the vertical, the mass
is a distance above the lowest point,

$h = L(1 - \cos\theta)$

$E = \frac{1}{2}mv^2 + mgL(1 - \cos\theta)$ □

(b) $U = mgL(1 - \cos\theta)$ an for small angles

$$U \cong mgL\left[1 - \left(1 - \frac{q^2}{2}\right)\right] = \frac{1}{2}mgL\,q^2$$

Also, since $qL = s$ and $w^2 = \frac{g}{L}$, we have $U = \dfrac{m\,\omega^2 s^2}{2}$ □

The simple pendulum is not a mass on a spring, but this
equation for potential energy applies also to a mass on a
spring: naming the spring extension as s, we have

$$U = \frac{1}{2}ks^2 = \frac{1}{2}m(\sqrt{k\,/\,m}\,)^2 s^2 = \frac{1}{2}m\,\omega^2 s^2.$$ One could define SHM

as motion in a well of potential energy proportional to the
squared excursion from equilibrium.

21.35 Problem: A pendulum of length 1 m is released from an initial angle of 15°. After 1000 s, its amplitude is reduced by friction to 5.5°. What is the value of $b/2m$?

Solution: According to Eq. 21.30, the amplitudes $\theta_0 = 15°$ and $\theta(t = 1000) = 5.5°$ must fit the pattern of exponential decay $A_t = A_0 e^{-bt/2\,m}$, where we use A_t to represent the amplitude at time t.

Then $\dfrac{A_t}{A_0} = \dfrac{Ae^{-bt/2m}}{A_0} = \dfrac{5.5}{15} = e^{-b(1000)/2m} = 0.367$

$e^{+b(1000)/2\,m} = 1/0.367 = 2.73$

$b(1000\ \text{s})/2\ m = \ln 2.73 = 1.00$

Then $\dfrac{b}{2m} = 1.00 \times 10^{-3}\,\text{s}^{-1}$ $\square$

21.39 Problem: A weight of 40 N is suspended from a spring with force constant 200 N/m. The system is undamped and is subjected to a harmonic force of frequency 10 Hz, resulting in a forced-motion amplitude of 2 cm. Determine the maximum value of the impressed force.

Solution: From Eq. 21.34, with $b = 0$ for undamped motion,

$A = \dfrac{F_0 / m}{\sqrt{\left(\omega^2 - \omega_0^2\right)^2}}$ Note $\omega = mg$ so $m = \omega/g = (40/9.8)$ kg

$\omega = 2\pi f = (20\pi s^{-1})$ $\omega_0^2 = \dfrac{k}{m} = \dfrac{200}{(40\ /\ 9.8)} = 49\ s^{-2}$

$F_0 = mA \left(\omega^2 - \omega_0^2\right)$

$F_0 = \left(\dfrac{40}{9.8}\right)(0.02)(3950 - 49) = 318\ \text{N}$ $\square$

21.43 **Problem:** Consider the circuit shown in Figure 21.18. Let $R = 7.6$ Ω, $l = 2.2$ mH, and $C = 1.8$ μF. (a) Calculate the frequency of the damped oscillation of the circuit. (b) What is the value of the critical resistance in the circuit?

Solution:

(a) $\omega_d = \sqrt{\dfrac{1}{LC} - \left(\dfrac{R}{2L}\right)^2} = \sqrt{\dfrac{1}{(2.2 \times 10^{-3})(1.8 \times 10^{-6})} - \left(\dfrac{7.6}{2(2.2 \times 10^{-3})}\right)^2}$

$\omega_d = 1.580 \times 10^4$ rad/s $\rightarrow f_d = \dfrac{\omega_d}{2\pi} = 2.51$ kHz $\square$

(b) $R_C = \sqrt{\dfrac{4L}{C}} = \sqrt{\dfrac{4(2.2 \times 10^{-3}\,\text{V} \cdot \text{s}/\text{ A})}{(1.8 \times 10^{-6}\,\text{A} \cdot \text{s}/\text{ V})}} = 69.9$ V/A $= 69.9$ Ω

21.47 **Problem:** An *RLC* circuit consists of a 150-Ω resistor, a 21-μF capacitor, and a 460-mH inductor, connected in series with a 120-V, 60-Hz power supply. (a) What is the phas angle between the current and the applied voltage? (b) Does the current or voltage reach its peak earlier?

Solution: The reactance of the inductor we symbolize as

$X_L = \omega L = 2\pi f L = 2\pi(60)(0.460) = 173.4$ Ω

The reactance of the capacitor is

$X_c = 1/\omega C = \dfrac{1}{2\pi f C} = \dfrac{1}{2\pi(60)(21 \times 10^{-6})} = 126.3$ Ω

(a) $\tan\phi = \dfrac{X_L - X_C}{R} = \dfrac{173.4\,\Omega - 126.3\,\Omega}{150\,\Omega} = 0.314$:

$\phi = 0.304$ rad $= 17.4°$ $\square$

(b) Since $X_L > X_C$, ϕ is positive, so V_{app} leads the current $\square$

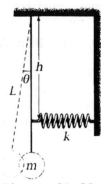

21.55 **Problem:** A pendulum of length L and mass M has a spring of force constant k connected to it at a distance of h below its point of suspension (Figure 21.55). Find the frequency of vibration of the system for small values of the amplitude (small θ). (Assume the vertical suspension of length L is rigid, but neglect its mass.)

Figure 21.55

Solution: For the pendulum (see sketch) we have $\Sigma\tau = I\alpha$ and $\dfrac{d^2\theta}{dt^2} = -\alpha$; the negative sign appears because positive θ is measured clockwise in the picture. Take torques around the point of suspension. $\Sigma\tau = MgL\sin\theta + kxh\cos\theta = I\alpha$

For small amplitude vibrations, use the approximations: $\sin\theta \approx \theta$, $\cos\theta \approx 1$, and $x \approx s = h\theta$.

Therefore, with $I = mL^2$,

$$\frac{d^2\theta}{dt^2} = -\left(\frac{MgL + kh^2}{I}\right)() = -\left(\frac{MgL + kh^2}{ML^2}\right)\theta$$

This is of the form $d^2\theta/dt^2 = -\omega^2\theta$ required for SHM, with angular frequency

$$\omega = \sqrt{\frac{MgL + kh^2}{ML^2}} = 2\pi f.$$

The ordinary frequency is $f = \dfrac{1}{2\pi}\sqrt{\dfrac{MgL + kh^2}{ML^2}}$ □

21.57 **Problem:** A flat plate, P, executes horizontal simple harmonic motion by sliding across a frictionless, surface with a frequency of $f = 1.5$ Hz. A block, B, rests on the plate, as shown in Figure 21.27, and the coefficient of static friction between the block and the plate is $\mu_s = 0.60$. What maximum amplitude of oscillation can the plate-block system have if the block is not to slip on the plate?

Solution: If the block does not slip, its motion is SHM with the same amplitude and frequency as the plate, and with its acceleration caused by the static friction force exerted on it by the plate. Think of the block when it is just ready to slip at a turning point in its motion:

$\Sigma F = ma$ becomes

$f_{max} = \mu_s n = \mu_s mg = ma_{max} = mA\omega^2$

Then $A = \mu_s g / \omega^2 = 0.6(9.8 \text{ m/s}^2)/(2\pi(1.5/\text{s}))^2$

$\quad\quad A = 6.62$ cm $\square$

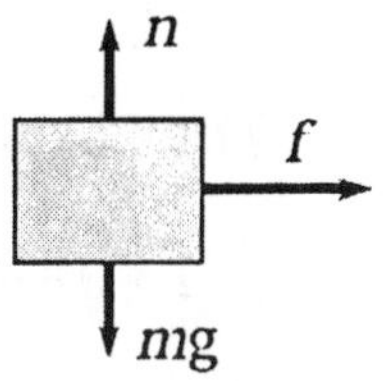

Figure 21.57

21.67 **Problem:** Consider a series *RLC* circuit with the following circuit parameters: $R = 200\ \Omega$, $L = 663$ mH, and $C = 26.5\ \mu F$. The applied voltage has an amplitude of 50 V and a frequency of 60 Hz. Find the following amplitudes: (a) The current, including its phase constant ϕ relative to the applied voltage; (b) the voltage V_R across the resistor and its phase relative to the current; (c) the voltage V_C across the capacitor and its phase relative to the current; and (d) the voltage V_L across the inductor and its phase relative to the current.

Solution: We identify $R = 200\ \Omega$, $L = 663$ mH, $C = 26.5\ \mu F$, $\omega = 377\ s^{-1}$, $V_{max} = 50$ V

Then $\omega L = 250\ \Omega$, $\left(\dfrac{1}{\omega C}\right) = 100W$

The impedance is

$$z = \sqrt{R^2 + (\omega L - 1/\omega C)^2} = \sqrt{(200\,\Omega)^2 + (250\,\Omega - 100\,\Omega)^2} = 250\ \Omega$$

(a) $I = \dfrac{V}{Z} = \dfrac{50V}{250\ \Omega} = 0.2$ A $\square$

$\phi = an^{-1}\left(\dfrac{X_L - X_R}{R}\right) = 36.8°$ $\square$, V leads I.

(b) $V_R = IR = 40$ V $\square$ at $\phi = 0°$

(c) The charge on the capacitor is related to the current by $I = dq/dt$

$q = \int I dt = \int I_m \sin(\omega t - \phi) dt = \int (I_m / \omega) \sin(\omega t - \phi)\omega dt$

$= -(I_m/\omega)\cos(\omega t - \phi)$

So the time-varying capacitor voltage is

$V = q/C = -(I_m/\omega C)\cos(\omega t - \phi)$ and its amplitude is

$V_C = I_m/\omega C = (0.2$ A$)(100\ \Omega) = 20.0$ V $\square$ at

$\phi = -90°$ (the current leads the voltage)

21.67 (Cont)

(d) The time-varying inductor voltage is

$$V = LdI/dt = L(d/dt)I_m\sin(\omega t - \phi)$$

$$= LI_m\omega \cos(\omega t - \phi), \text{ with amplitude}$$

$$V_L = I\omega L = 50 \text{ V } \square \text{ at } \phi = +90° \quad (V \text{ leads } I)$$

21.69 Problem: An object attached to the end of a spring vibrates with an amplitude of 20 cm. Find the position of the object at these times, 0, $T/8$, $T/4$, $3T/8$. $T/2$, $5T/8$, $3T/4$, $7T/8$, and T, where T is the period of vibration. Plot your results (position along the vertical axis and time along the horizontal axis).

Solution: In $x = A\cos(\omega t + \phi) = A\cos(2\pi f t + \phi)$

$x = A\cos(2\pi t/T + \phi)$, we arbitrarily choose $\phi = 0$ and substitute $A = 20$ cm and, sequentially, each given value for t. At $t = 0$, $x = 20$ cm cos 0 = 20 cm. At $t = T/8$,

$x = 20$ cm cos $(2\pi/8) = 20$ cm cos 45° = 14.1 cm. Similarly,

t	0	$T/8$	$T/4$	$3T/8$	$T/2$	$5T/8$	$3T/4$	$7T/8$	T
x, cm	20	14.1	0	-14.1	-20	-14.1	0	14.1	20

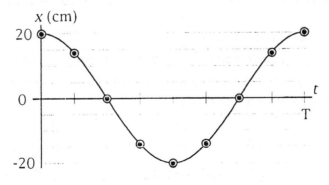

Figure 21.69

CHAPTER 22

QUESTIONS

4   **Question:** Can two pulses traveling in opposite directions on the same string reflect from each other? Explain.

Answer: No. Wave pulses will not reflect from each other; rather they pass right through each other. The wave is a pattern of passing-on of a disturbance and not a solid object.

5   **Question:** Harmonic waves are generated on a string under constant tension by a vibrating source. If the power delivered to the string is doubled, by what factor does the amplitude change? Does the wave speed change under these circumstances?

Answer: Power is always proportional to the square of amplitude, so if power doubles, amplitude increases by $\sqrt{2}$ times, by the factor 1.41. The wave speed does not depend on amplitude, but stays constant.

12   **Question:** In a longitudinal wave in a spring, the coils move back and forth in the direction the wave travels, does the speed of the wave depend on the maximum speed of each of the coils?

Answer: The wave speed does not depend on the speed of particle motion. The amplitude can be reduced or increased at will to make the particles' maximum speed small or large, all without any change in the speed of the wave.

16   **Question:**   As a result of a distant explosion, an observer senses
a ground tremor and then hears the explosion.   Explain.

Answer: Because it involves compression, the ground tremor can
be thought of as a sound wave moving through the earth as a
medium.   Rock is so much stiffer against compression than air
that the ground wave moves faster.   Starting together, the
vibration through the ground reaches the observer before the
sound in the air.

19   **Question:**   Suppose an observer and a source of sound are both at
rest, and a strong wind blows from a source toward the observer.
Describe the effects (if any) of the wind on (a) the observed
wavelength, (b) the observed frequency, and (c) the wave speed.

Answer: In the reference frame of the air - the observer is
moving toward the source at the wind speed through stationary
air, and the source is moving away from the observer with the
same speed.   In the Doppler frequency equation both the observer
speed and the source speed are counted as positive.   Then the
observed frequency $f' = f(v_{sound} + v_{wind})/(v_{sound} + v_{wind})$ is the
same as if no wind were blowing.   But the observer sees the sound
coming toward him at $(v_{sound} + v_{wind})$.   With this larger speed, he
attributes a larger wavelength $\lambda' = (v_{sound} + v_{wind})/f$ to the
wave.

Back in the earth frame of reference, we can say that the wind
motion spreads apart the crests of the sound wave traveling
toward the observer, just as would a less dense medium with a
higher sound speed.   But the frequency is left constant, just as
it would be by substituting a less dense medium.

Alternative Solution
(c) The wave speed is $v_{sound} + v_{wind}$
(b) The observed frequency is the same f as the source.
(a) The observed wavelength is $(v_{sound} + v_{wind})/f$

23   **Question:**   A sound wave travels in air at a frequency of 500 Hz.
      If part of the wave travels from the air into water, does its
      frequency change? Does its wavelength change? Justify your
      answers.   Note that the speed of sound in air is about 340 m/s,
      whereas the speed of sound in water is about 1500 m/s.

      Answer: There is no frequency change when a wave moves across a
      stationary interface between mediums.   Every crest in the first
      medium becomes just one crest in the second, so the number of
      cycles per second is the same on both sides.
      As the speed increases across the boundary from 340 m/s to
      1500 m/s, the wavelength steps up from (340 m/s)/(500/s) = 0.68 m
      to (1500 m/s)/(500/s) = 3 m.   The textbook Figures 25.9, 25.10,
      and 25.18 show how wavelength changes with speed, an effect
      called <u>refraction</u>.

CHAPTER 22

PROBLEMS

22.3 **Problem:** At $t = 0$, a transverse wave pulse in a wire is described by the function $y = \dfrac{6}{x^2 + 3}$ where x and y are in meters. Write the function $y(x,t)$ that describes this wave if it is traveling in the positive x direction with a speed of 4.5 m/s.

Solution: For definiteness, consider the point $x = 0$, $y = 2$ at $t = 0$. This is the center of the pulse. Take a time 4 s later. The center of the pulse must be $y = 2$ at (4 s)(4.5 m/s) = 18 m. For the whole pulse to have the same shape, the wave function must then be $y(x, 4s) = 6/((x - 18)^2 + 3)$. Now we can see that at any time the wave function is

$$y(x, t) = \frac{6}{(x - 4.5t)^2 + 3} \quad \square$$

22.9   **Problem:**   One wave pulse in a string is described by the

equation $y_1 = \dfrac{5}{(3x - 4t)^2 + 2}$.   A second wave pulse in the same

string is described by $y_2 = \dfrac{-5}{(3x + 4t - 6)^2 + 2}$

(a) In which direction does each pulse travel   (b) At what time
will the two waves exactly cancel everywhere?   (c) At what
point do the waves always cancel?

Solution:

(a)   A point of constant phase, constant $(3x - 4t)$, will be at
     larger x when t gets larger.   The $(x - vt)$ form of the
     expression means the wave y_1 moves to the right □.   In y_2,
     for $(3x + 4t - 6)$ to stay constant as t increases, x must
     decrease.   The wave is of the form $f(x + vt)$ and so moves
     to the left □ with speed $v = 4/3$.

(b)   We require $y_1 + y_2 = 0$:

$$\frac{5}{\left((3x - 4t)^2 + 2\right)} - \frac{5}{\left((3x + 4t - 6)^2 + 2\right)}$$

$$\frac{5}{(3x - 4t)^2 + 2} = \frac{+5}{(3x + 4t - 6)^2 + 2}$$

$$(3x - 4t)^2 = (3x + 4t - 6)^2$$

$$3x - 4t = \pm(3x + 4t - 6)$$

The positive root allows $3x - 4t = 3x + 4t - 6$.   This
will be true for all x at the special time $8t = 6$,
$t = 0.750$ s □.   At this time the waves cancel
everywhere.

(c)   The negative root allows $3x - 4t = -3x - 4t + 6$.   This
will be true for all t at the special point $6x = 6$,
$x = 1.00$ m □.   There the waves cancel always.

22.13 **Problem:** Transverse waves travel with a speed of 20 m/s in a string that is under a tension of 6 N. What tension is required for a wave speed of 30 m/s in the same string?

Solution: If the linear density remains constant, $v_1 = \sqrt{F_1 / \mu}$ and $v_2 = \sqrt{F_2 / \mu}$. Dividing, $v_2/v_1 = \sqrt{F_2 / F_1}$ and $F_2 = (v_2/v_1)^2 F_1 = (30/20)^2 6$ N $= 13.5$ N $\square$

22.19 **Problem:** A transverse wave moving along a string in the positive x direction with a speed of 200 m/s has an amplitude of 0.7 mm and a wavelength of 20 cm. Determine (in SI units) the values of A, k, and ω in the equation describing the wave: $y = A \sin (kx - \omega t)$.

Solution: $A = 0.7$ mm $\square$

$$k = \frac{2\pi}{\lambda} = \frac{2\pi}{0.2 \text{ m}} \left(\frac{1 \text{ rad}}{1}\right) = 31.4 \text{ rad/m } \square$$

$$\omega = 2pf = 2\pi \frac{v}{\lambda} = \frac{2\pi(200 \text{ m/ s})}{(0.2 \text{ m})} = 6280 \text{ rad/s } \square$$

So the wave function is

$A \sin (kx - \omega t) = 0.7$ mm $\sin (31.4$ $x/m - 6280$ $t/s)$

22.25 **Problem:** A wave form is described by

y = (2.0 cm) sin (kx - ωt)

k = 3.1 rad/cm

ω = 9.3 rad/s

where x is the position along the wave form (in meters) and t is the time (in seconds). Determine the amplitude, wave number, wavelength, angular frequency, and speed of the wave.

Solution: We compare y = A sin(kx - ωt) with the given with ψ = 2 cm sin(2.11 rad x/m - 3.62 rad t/s) to identify that the wave disturbance has been renamed from y to ψ:

The amplitude is A = 2.00 cm □

The wave number k = 2.11 rad/m □

The wavelength is $\lambda = \dfrac{2\pi}{k}$ = 2.98 m □

The angular frequency w = 3.62 rad/s □

And the speed is $v = f\lambda = \dfrac{\omega}{2\pi} \cdot \dfrac{2\pi}{k} = \dfrac{\omega}{k} = \dfrac{3.62}{2.11}$ = 1.72 m/s □

22.29 Problem: Harmonic waves 5 cm in amplitude are to be transmitted along a string that has a linear density of 4×10^{-2} kg/m. If the maximum power delivered by the source is 300 W and the string is under a tension of 100 N, what is the highest vibrational frequency at which the source can operate?

Solution: $A = 5$ cm $= 5 \times 10^{-2}$ m; $m = 4 \times 10^{-2}$ kg/m; $P = 300$ W; $F = 100$ N

Therefore, $v = \sqrt{\dfrac{F}{\mu}} = 50$ m/s is the wave speed.

In $P = \frac{1}{2}mw2A^2v$ maximum power corresponds to highest ω:

$$\omega^2 = \frac{2P}{\mu A^2 v} = \frac{2(300)}{(4 \times 10^{-2})(5 \times 10^{-2})^2(50)}$$

$\omega = 346.4$ rad/s

$f = \dfrac{\omega}{2\pi} = 55.1$ Hz $\square$

22.37 Problem: Write an expression that describes the pressure variation as a function of position and time for a harmonic sound wave in air if its wavelength is 0.1 m and $\Delta P_m = 0.2$ Pa.

Solution: We write the wave function
$\Delta r = \Delta P_m \sin(kx - \omega t)$ by identifying

$$k = \frac{2\pi}{\lambda} = \frac{2\pi}{(0.1 \text{ m})} = 62.8 \text{ m}^{-1}$$

$$w = \frac{2\pi v}{\lambda} = \frac{2\pi(343 \text{ m/ s})}{(0.1 \text{ m})} = 2.16 \times 10^4 \text{ s}^{-1}$$

Therefore, $\Delta P = (0.2)\sin[62.8x - 2.16 \times 10^4 \, t]$ Pa $\square$

22.41 **Problem:** A tuning fork vibrating at 512 Hz falls from rest and accelerates at 9.80 m/s^2. How far below the point of release is the tuning fork when waves of frequency 485 Hz reach the release point? Take the speed of sound in air to be 340 m/s.

Solution: As the fork speeds up, the frequency heard back at the release point changes. Think of one particular wave crest radiated by the fork at time t_{fall} after its motion began, when its speed is $v_{source} = 9.8$ m/s^2 t_{fall}. This is received as one crest in sound of observed frequency 485 Hz. For it we have:

$$f_1 = f \left(\frac{v}{v + v_s} \right)$$

$$485 = 512 \left(\frac{340}{340 + 9.8\ T_{fall}} \right)$$

$$485(340) + (485)(9.8 t_f) = (512)(340)$$

$$t_f = \left(\frac{512 - 485}{485} \right) \frac{340}{9.8} = 1.93 \text{ s.}$$

In this time the fork fell a distance $d_1 = \frac{1}{2} g t_f^2 = 18.278$ m. Now our crest will take time t_{return} to get back up to the starting point:

$$t_{return} = \frac{18.3}{340} = 0.05376 \text{ s.}$$

The fork falls some more while sound returns

$$t_{total\ fall} = 1.93 \text{ s} + 0.05376 \text{ s} = 1.9852 \text{ s}$$

$$d_{total} = \frac{1}{2} g t_{total}^2 = 19.3 \text{ m} \ \square$$

22.49 **Problem:** A supersonic jet traveling at Mach 3 at an altitude of 20,000 m is directly overhead at time $t = 0$, as in Figure 22.49. (a) How long will it be before the observer encounters the shock waves? (b) Where will the plane be when it is finally heard? (Assume the speed of sound in air is 335 m/s.)

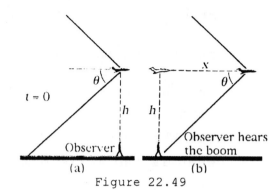

Figure 22.49

Solution: We can solve part (b) first. The plane will have moved forward by the distance x shown.

$$\sin \theta = \frac{v}{v\,3} = \frac{1}{3} \qquad \theta = 19.5°$$

$$\tan \theta = \frac{h}{x} \qquad x = \frac{h}{\tan \theta}$$

$$x = \frac{20000 \text{ m}}{\tan 19.5°} = 5.66 \times 10^4 \text{ m} = 56.6 \text{ km } \square$$

Now (part a) it takes the plane

$$t = \frac{x}{v\,s} = \frac{5.66 \times 10^4 \text{ m}}{3 \times 335 \text{ m/ s}} = 56.3 \; \square \text{ to travel this distance.}$$

22.51 Problem: The wave function for a transverse wave on a taut string is (in SI units): $y(x,t) = (0.35 \text{ m}) \sin\left(10\pi - 3\pi x + \dfrac{\pi}{4}\right)$

(a) What is the velocity of the wave? (Give the speed and the direction.) (b) What is the displacement at $t = 0$, $x = 0.2$ m? (c) What are the wavelength and frequency of the wave? (d) What is the maximum magnitude of the transverse velocity of the string?

Solution: We compare $y = A \sin (kx - \omega t + f)$ with

$y = (0.35 \text{ m}) \sin (10\pi t - 3\pi x + \pi/4)$

$y = -(0.35 \text{ m}) \sin (3\pi x - 10\pi t - \pi/4)$

to see $k = 3\pi/\text{m}$ and $\omega = 10\pi/\text{s}$

(a) The speed is $v = fl = 2\pi f \dfrac{\lambda}{2\pi} = \dfrac{\omega}{k} = \dfrac{10\pi \ / \ \text{s}}{3\pi \ / \ \text{m}} = 3.33$ m/s

in the positive x-direction

(b) $y(0.1 \text{ m}, 0) = (0.35 \text{ m}) \sin(0 - 3\pi(0.1) + \pi/4)$

$y = (0.35 \text{ m}) \sin(-0.157) = (0.35 \text{ m}) \sin(-0.157 \text{ rad})$

$y = (0.35 \text{ m}) \sin(-0.157 \text{ rad } 360°/2p \text{ rad})$

$y = (0.35 \text{ m}) \sin(-9°) = -5.48 \text{ cm } \square$

Note well that when you take the sine of a quantity with no units its units are not degrees, but radians.

(c) $k = \dfrac{2\pi}{\lambda} = 3\pi$ $\lambda = 0.667$ m $\square$

$\omega = 2\pi f = 10\pi$ $f = 5.00$ Hz $\square$

(d) $v_y = \dfrac{dy}{dt} = (0.35)(10\pi) \cos\left(10\pi t - 3\pi x + \dfrac{\pi}{4}\right)$

$v_{y,\text{max}} = (10\pi)(0.35) = 11.0$ m/s

Note how different this maximum particle speed is from the wave speed.

22.55 **Problem:** An earthquake on the ocean floor in the Gulf of Alaska induces a *tsunami* (sometimes called a "tidal wave") that reaches Hilo, Hawaii, 4450 km distant, in a time of 9 h 30 min. Tsunamis have enormous wavelengths (100 to 200 kg), and for such waves the propagation speed is $v \approx \sqrt{g\overline{d}}$, where $\overline{d}$ is the average depth of water. From the information given fink the aver wave speed and the average ocean depth between Alaska and Hawaii. (This method was used in 1856 to estimate the average depth of the Pacific Ocean long before soundings were made to give a direct determine.)

Solution: $v = \dfrac{4450 \text{ km}}{9.5 \text{ h}} = 468 \text{ km/h} = 130 \text{ m/s}$ □

$$d = \frac{v^2}{g} = \frac{(130 \text{ m/ s})^2}{9.8 \text{ m/ s}^2} = 1.73 \text{ km} \quad \square$$

22.61 **Problem:** In order to be able to determine her speed, a skydiver carries a tone generator. A friend on the ground at the
landing site has equipment for receiving and analyzing sound
waves. While the skydiver is failing at terminal speed, her
tone generator emits a steady tone of 1800 Hz. (Assume that
the air is calm and that the sound speed is 343 m/s,
independent of altitude.) (a) If her friend on the ground
(directly beneath the skydiver) receives waves of frequency
2150 Hz, what is the skydiver's speed of descent? (b) If the
skydiver were also carrying sound-receiving equipment sensitive
enough to detect waves reflected from the ground, what
frequency would be the relationship.

Solution: Call f_e = 1800 Hz the emitted frequency and f_g =
2150 Hz the frequency of wave crests reaching the ground. Then

(a) $f_g = f_e \dfrac{v}{(v - v_{diver})}$, so $1 - \dfrac{v_{driver}}{v} = \dfrac{f_e}{f_g}$

$\therefore \; v_{diver} = v \left(1 - \dfrac{f_e}{f_g} \right)$ with v = 343 m/s, f_e = 1800 Hz and

f_g = 2150 Hz, we find $v_{diver} = 343 \left(1 - \dfrac{1800}{2150} \right)$ = 55.8 m/s □

(b)   Now the ground acts as a source, reflecting wave crests
with the frequency f_g at which they reach the ground. The
skydiver is an observer moving into the waves, to receive
frequency

$$f_{rec} = f_g \dfrac{(v + v_{diver})}{v} - f_{rec} = f_e \left[\dfrac{v}{(v - v_{diver})} \right] \dfrac{(v + v_{diver})}{v}$$

so $f_{rec} = 1800 \dfrac{(343 - 55.8)}{(343 - 55.8)}$ = 2500 Hz □

CHAPTER 23

QUESTIONS

3   **Question:** What are the conditions necessary for the production of standing waves with well-defined nodes?

Answer: Two identical traveling waves moving in opposite directions add to give a standing wave. You can set up a standing wave by reflecting one periodic wave back upon itself from a single efficient reflecting surface. If you choose to use more than one reflector, confining the wave to a restricted region of space, then they need not be efficient reflectors; but then the frequency must be one of a certain set of special values, the resonance frequencies: the frequencies of free vibrations of the medium in that restricted space.

13   **Question:** When the base of a vibrating tuning fork is placed against a chalkboard, the sound becomes louder due to resonance. How does this affect the length of time the fork vibrates? Does this agree with conservation of energy?

Answer: Instead of radiating sound very softly into the surrounding air, the tuning fork makes the chalkboard vibrate; with its large area it radiates sound into the air with higher power. So it drains away the fork's energy of vibration faster and the fork stops vibrating sooner. This is also an example of energy conservation, as the energy of vibration of the fork is transferred -- via the blackboard -- into energy of vibration of the air (sound).

18   **Question:** Despite a reasonably steady hand, a certain person often spills his coffee when carrying it to his seat. Discuss resonance as a possible cause of this difficulty, and devise a means for solving the problem.

Answer: Walking makes the person's hand vibrate a little. If the frequency of this motion equals the natural frequency of coffee sloshing from side to side in the cup, then a large-amplitude vibration of the coffee will build up in resonance. To get off resonance and back to the normal case of a small-amplitude disturbance producing a small-amplitude result, the person can walk faster, walk slower, or get a cup of larger or smaller diameter.

19   **Question:** A soft-drink bottle resonates as air is blown across the top. What happens to the resonant frequency as the level of fluid in the bottle decreases?

Answer: Blowing across the top will make a noise to excite resonance of the air inside in its simplest standing-wave vibration possibility, with an antinode at the month and a node at the liquid surface. The diameter of the air column inside the bottle is not uniform, so the wavelength of the broadcast sound may not be just four times the air-column length. But still drinking more of the soft drink will increase the length of the air column to increase the wavelength and lower the frequency.

CHAPTER 23

PROBLEMS

23.1 **Problem:** Two harmonic waves are described by

$y_1 = (0.5 \text{ m}) \sin [\pi(4x - 1200t)]$

$y_2 = (0.5 \text{ m}) \sin [\pi(4x - 1200t - 0.25)]$ where x, y_1, and y_2 are in meters and t is in seconds. (a) What is the amplitude of the resultant wave? (b) What is the frequency of the resultant wave?

Solution: We can represent the waves symbolically as

$y_1 = A_0 \sin(kx - \omega t)$ and

$y_2 = A_0 \sin(kx - \omega t - f)$

with $A_0 = 0.5$ m, $\omega = 1200\pi/s$, and $\phi = 0.25\pi$.

The resultant wave function has the form

$y = y_1 + y_2 = 2A_0 \cos (\phi/2) \sin (kx - \omega t - \phi/2)$, with amplitude

(a) $A = 2A_0 \cos \left(\dfrac{\phi}{2}\right) = 2(.5) \cos \left[\dfrac{(\pi / 4)}{2}\right] = 0.924$ m $\square$ and frequency

(b) $f = \dfrac{\omega}{2\pi} = \dfrac{1200\pi}{2\pi} = 600$ Hz $\square$

23.5   **Problem:**   Two identical sound sources are located along the y axis.   Source S_1 is located at $(0,\ 0.1)$ m, and source S_2 is located at $(0,\ -0.1)$ m.   The two sources radiate isotopically at a frequency of 1715 Hz, and the amplitude of each wave separately is assumed to be A.   A listener is located along the y axis a distance of 5 m from source S_1.   (a) What is the phase difference between the sound waves at the position of the listener?   (b) What is the amplitude of the resultant wave at the location of the listener?   (Use $v = 343$ m/s.)

   Solution: (a) To reach the listener the wave from source S_2 must travel an extra distance

   $$\Delta r = r_2 - r_1 = 5.2 - 5.0 = 0.2 \text{ m.}$$

   So the phase difference is

   $$\Delta\phi = \left(\frac{2\pi}{\lambda}\right)\Delta r, \text{ where } \lambda = \frac{v}{f} = \frac{343}{1715} = 0.2 \text{ m}$$

   $$\Delta\phi = \left(\frac{2\pi}{0.2}\right)(0.2) = 6.28 \text{ rad } \square$$

   In a word, the waves are <u>in phase</u> to interfere constructively and produce a resultant wave with amplitude.

   (b) $A' = \left| 2A \cos\left(\frac{\phi}{2}\right) \right| = 2.00 \ A \ \square$

23.9 **Problem:** Two waves in a long string are

$$y_1 = (0.015 \text{ m})\cos\left(\frac{x}{2} - 40t\right), \quad y_2 = (0.015 \text{ m})\cos\left(\frac{x}{2} + 40t\right) \text{ where}$$

the y's and x are in meters and t is in seconds. (a) Determine the positions of the nodes of the resulting standing wave. (b) What is the maximum displacement at the position $x = 0.4$ m?

Solution: According to $\cos(a \pm b) = \cos a \cos b \mp \sin a \sin b$, the two waves add to $y = y_1 + y_2 = (0.03 \text{ m})\cos(x/2)\cos 40t$.

(a) The nodes occur where $y = 0$, where $\cos(x/2) = 0$,

at $x/2 = (2n + 1)\pi/2$ for $n = 1, 2, 3, \ldots$;

so $x = (2n + 1)\ \pi = \pi, 3\pi, 5\pi, \ldots$

(b) $y_{\max} = 0.03\cos\left(\frac{0.4}{2}\right) = 0.0294 \text{ m } \square$

23.15 **Problem:** The ship in Figure 23.15 travels along a straight line that is parallel to the shore and 600 m from the shore. The ship's radio receives simultaneous signals of the same frequency from antennas A and B. The signals interfere constructively at point C, which is equidistant from A

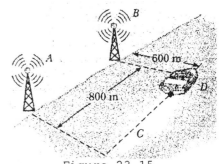

Figure 23.15

and B. The signal goes through the first minimum at point D. Determine the wavelength of the radio waves.

Solution: At D, the ship is 600 m from B and distant from A by $\sqrt{600^2 + 800^2}$ m $= 1000$ m. Thus, the extra distance of travel for the wave from source A is 400 m. Since D is the first minimum, we have $\dfrac{\lambda}{2} = 400$ m $\quad \lambda = 800$ m $\square$

23.19 **Problem:** A cello A-string vibrates in its fundamental mode with a frequency of 220 vibrations/s. The vibrating segment is 70 cm long and has a mass of 1.2 g. (a) Find the tension in the string. (b) Determine the frequency of the harmonic that causes the string to vibrate in three segments.

Solution: d_{NN} = 0.7 m for the simplest vibration possibility. Then λ = 1.4 m

$$f\lambda = v = 308 \text{ m/s} = \sqrt{\frac{F}{1.2 \times 10^{-3}/0.7}}$$

(a) F = 163 N □

(b)   For the thud vibration state the tension, linear density, and speed are the same. The (NANANAN) standing wave has

d_{NN} = 0.7 m/3; λ = 2(0.7 m)/3 = 0.467 m; and

f_3 = v/λ = (308 m/s)/(0.467 m) = 660 Hz □

23.25 **Problem:** Standing-wave vibrations are set up in a crystal goblet with two nodes and two antinodes equally spaced around the 20-cm circumference of its rim. If transverse waves move around the glass at 900 m/s, and opera singer would have to produce a high harmonic with what frequency in order to shatter the glass with a resonant vibration?

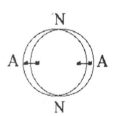

Figure 23.25

Solution: The distance between adjacent nodes is half the circumference $d_{N \text{ to } N}$ = $d_{A \text{ to } A}$ = $\frac{\lambda}{2}$ = 10 cm so λ = 20 cm and

$$f = \frac{v}{\lambda} = \frac{900 \text{ m/ s}}{0.2 \text{ m}} = 4500 \text{ Hz.} \ \square$$ The singer must match this

frequency quite precisely for some interval of time to feed enough energy into the glass to crack it.

23.31 **Problem:** A glass tube (open at both ends) of length L is positioned near an audio speaker of frequency f = 0.68 kHz. For what values of L will the tube resonate with the speaker?

Solution: The wave in the air will always have frequency f = 680 Hz and wavelength $\lambda = v/f$ = (344 m/s)/(680/s) = 0.506 m. So the standing wave in the tube will always have $d_{A \text{ to } A}$ = $\lambda/2$ = 0.253 m. The air column can resonate in its simplest state (ANA) with length $L = d_{A \text{ to } A}$ = 0.253 m. It can resonate in vibration possibility (ANANA) with $L = 2d_{A \text{ to } A}$ = 0.506 m. The speaker can make it vibrate as (ANANANA) with $L = 3d_{A \text{ to } A}$ = 0.759 m; or in general the tube length can be $L = n(0.253 \text{ m})$ with n = 1, 2, 3,...

23.37 **Problem:** In certain ranges of a piano keyboard, two or more strings are tuned to the same note to provide extra loudness. For example, two strings are tuned to 110 Hz. If one of the two slips from its normal tension, 600 N, to 540 N, what beat frequency will be heard when the two strings are struck simultaneously?

Solution: The string of length L vibrates as (NAN) with the wavelength of the string wave equal to $2L$. The speed is $v = \sqrt{F / \mu}$ and its frequency is $f = v/\lambda = \sqrt{F / \mu} / 2L$. Thus the frequency is proportional to the square root of tension. Call the new frequency $f_{new} = \sqrt{F_{new} / \mu} / 2L$. Dividing gives $f_{new}/f = \sqrt{F_{new} / F}$ and the second string vibrates at

$$f_{new} = 110\sqrt{\frac{540}{600}} = 104.4 \text{ Hz.}$$

The beat frequency is the difference,

$$f_b = |f_1 - f_2| = 110 - 104.4 = 5.6 \text{ Hz} \quad \square$$

23.45 Problem: If two adjacent natural frequencies of an organ pipe
are determined to be 0.55 Hz and 0.65 kHz, calculate the
fundamental frequency and length of this pipe.
(Use v = 340 m/s.)

Solution: The spacing of harmonics in frequency is numerically
equal, so we can count down to the fundamental frequency:
650 Hz, 550 Hz, 450 Hz, 350 Hz, 250 Hz, 150 Hz, 50 Hz.   The
fundamental is at 50.0 Hz $\square$, with only odd-numbered harmonics,
the pipe must be closed at one end. The wavelength of the (N-A)
fundamental vibration is λ = v/f = (340 m/s)/(50/s) = 6.8 m, so
the pipe length is d_{NA} = $\lambda/4$ = 1.70 m $\square$

23.49 Problem: Two train whistles have identical frequencies of
180 Hz.   When one train is at rest in the station, sounding its
whistle, a beat frequency of 2 Hz is heard from a moving train.
What two possible speeds and directions can the moving train
have?

Solution: If the second train is moving away from station, the
frequency is depressed:

f' = 180 - 2 = 178 Hz $178 = 180 \dfrac{(343)}{(343 + v)}$

Solving for v gives $v = \dfrac{(2)(343)}{178}$

Therefore, v = 3.85 m/s away from station $\square$

Moving towards the station, the frequency is enhanced:

f' = 180 + 2 = 182 Hz $182 = 180 \dfrac{(343)}{(343 - v)}$

Solving for v gives $v = \dfrac{(2)(343)}{182}$

Therefore, v = 3.77 m/s towards the station $\square$

23.55 **Problem:** Two wires are welded together. They are of the same material, but one is twice the diameter of the other. The wires are subjected to a tension of 4.6 N. The thin wire has a length of 40 cm and a linear mass density of 2 g/m. The combination is fixed at both ends and vibrated in such a way that two antinodes are present, with the central node being right at the weld. (a) What is the frequency of vibration? (b) How long is the thick wire?

Solution:

(a) Since the first node is at the weld, the wave-length in the thin wire is $2L$ or 80 cm. The frequency and tension are the same in both sections, so

$$f = \frac{1}{2L} \sqrt{\frac{F}{\mu}} = \frac{1}{2(0.4)} \sqrt{\frac{4.6}{0.002}} = 60 \text{ Hz } \square$$

(b) As the thick wire is twice the diameter, its cross-sectional area is four times larger, and the linear density is four times that of the thin wire.

$$m' = 8g/m \text{ so } L' = \frac{1}{2f} \sqrt{\frac{F}{\mu'}}$$

$$L' = \left[\frac{1}{(2)(60)}\right] \sqrt{\left[\frac{(4.6)}{(0.008)}\right]} = 20 \text{ cm } \square \text{ half the length of the}$$

thin wire.

23.57 **Problem:** A standing wave is set up in a string of variable length and tension by a vibrator of variable frequency. When the vibrator has a frequency of f in a string of length L and tension F, n antinodes are set up in the string. (a) If the length of the string is doubled, by what factor should the frequency be changed to get the same number of antinodes? (b) If the frequency and length are held constant, what tension will produce $n + 1$ antinodes? (c) If the frequency is tripled and the length halved, by what factor should the tension be changed to get twice as many antinodes?

Solution:

(a) $f = \dfrac{n}{2L}\sqrt{\dfrac{F}{\mu}}$ so $\dfrac{f'}{f} = \dfrac{L}{L'} = \dfrac{L}{2L} = \dfrac{1}{2}$ $\square$

The frequency should be halved to get the same number of antinodes for twice the length.

(b) $\dfrac{n'}{n} = \sqrt{\dfrac{F}{F'}}$ so $\dfrac{F'}{F} = \left(\dfrac{n}{n'}\right)^2 = \left[\dfrac{n}{(n+1)}\right]^2$

The tension must be $F' = \left[\dfrac{n}{(n+1)}\right]^2 F$

(c) $\dfrac{f'}{f} = \dfrac{n'}{n}\dfrac{L}{L'}\sqrt{\dfrac{F'}{F}}$ so $\dfrac{F'}{F} = \left(\dfrac{f'n'L}{f'nL}\right)^2$

$\dfrac{F'}{F} = \left(\dfrac{3}{2\cdot 2}\right)^2 = \dfrac{9}{16}$ $\square$

CHAPTER 24

QUESTIONS

10   **Question:** Suppose a creature from another planet had eyes that were sensitive to infrared radiation. Describe what he would see if he looked around the room you are now in. That is, what would be bright and what would be dim?

Answer: Light bulbs and the toaster glow brightly in the infrared. Somewhat fainter are the back of the refrigerator and the back of the television set, while the TV screen is dark. The pipes under the sink show the same weak glow as the walls until you turn on the faucets. Then the pipe on the right gets darker while that on the left develops a rich glow that quickly runs up along its length. The food on your plate shines; so does human skin, the same color for all races. Clothing is dark as a rule, but your seat glows like a monkey's rump when you get up from a chair, and you leave a patch of the same glow behind on the chair. Your face shows you are lit from within, like a jack-o-lantern: your nostrils and openings of your ear canals are bright; brighter still are the pupils of your eyes.

13   **Question:**   Radio stations often advertise "instant news." If what they mean is that you hear the news at the instant they speak it, is their claim true? About how long would it take for a message to travel across this country by radio waves, assuming that these waves could travel this great distance and still be detected?

Answer: Radio waves move at the speed of light. They can travel around the curved surface of the Earth, bouncing between the ground and the ionosphere, which has an altitude small compared to the radius of the Earth. The distance across the lower forty-eight United States is about 5000 km, requiring time $(5 \times 10^6 \text{ m})/(3 \times 10^8 \text{ m/s}) \sim 10^{-2} \text{ s }\square$. To go halfway around the Earth takes only 7×10^{-2} s. A speech can be heard on the other side of the world before it is heard in the back of the room.

14   **Question:**   Light from the Sun takes approximately $8\tfrac{1}{3}$ minutes to reach Earth. During this time the Earth has continued to move in its orbit around the Sun. How far is the actual location of the Sun from its image in the sky?

Answer: The sun's angular speed in our sky is our rate of rotation, $360°/24$ h $= 15°/$h. In $8\tfrac{1}{3}$ minutes it moves west by $\theta = \omega t = (15°/\text{h})(\text{h}/60 \text{ min})(8.3 \text{ min}) = 2.1°$. This is about four times the angular diameter of the sun.

CHAPTER 24

PROBLEMS

24.3   **Problem:** The magnetic field amplitude of an electromagnetic wave is 5.4×10^{-7}T. Calculate the electric field amplitude if the wave is traveling (a) in free space; (b) in a nonmagnetic medium in which the speed of the wave is $0.8c$.

Solution:

(a)   The fields in an electromagnetic wave are related by

$$\frac{E}{B} = c$$

$$\frac{E}{5.4 \times 10^{-7}} = 3 \times 10^{8}$$

$$E = 162 \text{ V/m} \ \square$$

(b) $\dfrac{E}{5.4 \times 10^{-7}} = 0.8(3 \times 10^{8})$

$$E = 130 \text{ V/m} \ \square$$

24.7   Problem: The electric field in an electromagnetic wave, in SI units, is described by $E_y = 100 \sin(10^7 x - \omega t)$. Find (a) the amplitude of the corresponding magnetic wave, (b) the wavelength, λ, and (c) the frequency, f.

Solution: (a) $B = \dfrac{E}{c} = \dfrac{100 \ \text{V/m}}{3 \times 10^8 \ \text{m/s}} = 3.33 \times 10^{-7} \ \text{T}$ $\square$

(b)   We compare the given wave function with

$$y = A \sin(kx - \omega t) \quad \text{to see} \quad k = 10^7 / \text{m}. \quad \text{Then}$$

$$\lambda = \frac{2\pi}{k} = \frac{2\pi}{10^7 \ \text{m}^{-1}} = 6.28 \times 10^{-7} \ \text{m} \ \square$$

(c) $f = \dfrac{c}{\lambda} = \dfrac{3 \times 10^8 \ \text{m/s}}{6.28 \times 10^{-7} \ \text{m}} = 4.78 \times 10^{14} \ \text{Hz}$ $\square$

24.9   Problem: If the coil in the resonant circuit of a radio has an inductance of 2 μH, what range of values must the tuning capacitor have in order to cover the complete range of FM frequencies?   (The FM range of frequencies is 99 MHz to 108 MHz.)

Solution: For the FM band, the frequencies range from a low of 88 MHz to a high of 108 MHz.   Thus from $f_0 = \dfrac{1}{2\pi\sqrt{LC}}$, we have

$$C = \frac{1}{4\pi^2 L f_0^2} \quad \text{which, with } L = 2 \times 10^{-6} \ \text{H, becomes}$$

$$C = \frac{1}{(7.90 \times 10^{-5} \ \text{Hs}) f_0^2}$$

(a) for a frequency of 108 MHz, $C = 1.09 \times 10^{-12}$ F = 1.09 pF $\square$

(b) for a frequency of 88 MHz, $C = 1.64 \times 10^{-12}$ F = 1.64 pF $\square$

24.13 **Problem:** A television set uses a dipole receiving antenna for VHF channels and a loop antenna for UHF channels. The UHF antenna produces a voltage from the changing *magnetic* flux through the loop. (a) Using Faraday's law, derive an expression for the amplitude of the voltage that appears in a single-turn circular loop antenna with a radius of r. The TV station broadcasts a signal with a frequency of f, and the signal has an electric field amplitude of E_{max} and a magnetic field amplitude of B_{max} at the receiving antenna's location. (b) If the electric field in the signal points vertically, what should be the orientation of the loop for best reception?

Solution: We suppose the diameter of the loop is smaller than the wavelength, so we can approximate the magnetic field as uniform over the area of the loop while it oscillates in time as $B = B_m \cos \omega t$. The induced voltage is

$$\varepsilon = - \frac{d \, \Phi_B}{dt} = - \frac{d}{dt} (BA \cos \theta) = - A \frac{d}{dt} (B_m \cos \omega t \cos \theta)$$

$$\varepsilon = AB_m \omega (\sin \omega t \cos \theta$$

$$\varepsilon(t) = 2\pi f B_m A \sin | \pi f t \cos \theta = 2\pi^2 r^2 f B_m \cos \theta \sin 2\pi f t.$$

(a) The amplitude of this emf is $\theta_m = 2\pi^2 r^2 f B_m \cos \theta$ where θ is the angle between the magnetic field and the normal to the loop.

(b) If **E** is vertical, then **B** is horizontal, so the plane of the loop should be vertical, and the plane should point toward the transmitter. This will make $\theta = 0°$, so $\cos \theta$ takes on its maximum value.

24.15 **Problem:** Two radio-transmitting antennas are separated by half the broadcast wavelength and are driven in phase with each other. (a) In which direction is the strongest signal radiated? (b) In which direction is the weakest signal radiated?

Solution: Along the perpendicular bisector of the line joining the antennas, you are equally distant from both. They oscillate in phase, so along this line you receive the two signals in phase. They interfere constructively to produce the strongest signal. Along the extended line joining the sources, the wave from the farther antenna must travel one-half wavelength farther to reach you, so you receive the waves 180° out of phase. They interfere destructively to produce the weakest signal.

24.23 **Problem:** The filament of an incandescent lamp has a 150-W resistance, and carries a *dc* current of 1 A. The filament is 8 cm long and 0.9 mm in radius. (a) Calculate the Poynting vector at the surface of the filament. (b) Find the magnitude of the electric and magnetic fields at the surface of the filament.

Solution: The rate at which the resistor converts electromagnetic energy into heat is $P = I^2 R$ = 150 W; The surface area is $A = 2\pi r L = 2\pi (0.9 \times 10^{-3} \text{ m})(0.08 \text{ m}) = 4.52 \times 10^{-4} \text{ m}^2$. The Poynting vector is

$$S = \frac{P}{A} = 3.32 \times 10^5 \text{ W/m}^2 \ \square \ \text{(points radially inward)}.$$

(b) $B = \mu_0 \dfrac{I}{2\pi r} = \dfrac{\mu_0 (1)}{2\pi\left(0.9 \times 10^{-3}\right)} = 2.22 \times 10^{-4} \text{ T} \ \square$

$\quad E = \dfrac{\Delta V}{\Delta x} = \dfrac{IR}{L} = \dfrac{150}{0.08} \dfrac{\text{V}}{\text{m}} = 1880 \text{ V/m} \ \square$

Note: $S = \dfrac{EB}{\mu_0} = 3.32 \times 10^5 \text{ W/ m}^2$

24.29 **Problem:** A 15-mW helium-neon laster (λ = 632.8 nm) emits a beam of circular cross-section whose diameter is 2 mm. Assume the beam has a uniform intensity over its cross section. (a) Find the maximum electric field in the beam. (b) What total energy is contained in a 1-m length of the beam? (c) Find the momentum carried by a 1-m length of the beam.

Solution: The intensity of the light is the average magnitude of the Poynting vector:

$$I = \frac{P}{\pi r^2} = \frac{E_m^2}{2 \mu_0 c}$$

Then the maximum electric field is

(a) $E_m = \sqrt{\dfrac{P \cdot 2 \mu_0 c}{\pi r^2}}$ = 1.90 x 10^3 N/C $\square$

(b) The power being 15 mW means that 15 mJ passes through a cross section of the beam in one second. This energy is uniformly spread through a beam length of 3 x 10^8 m, since that is how far the front end of the energy travels in one second. Thus the energy ir. just a one-meter length is

$$\frac{15 \times 10^{-3} \text{ J/ s}}{3 \times 10^8 \text{ m/ s}} = 5 \times 10^{-11} \text{ J/m}$$

(c) $p = \dfrac{U}{c} = \dfrac{5 \times 10^{-11}}{3 \times 10^8}$ = 1.67 x 10^{-19} kg · m/s $\square$

24.37   **Problem:** A radio receiver is located
200 km from a transmitter. A signal
travels to the receiver via a direct
line-of-sight path and also via a wave
from an ionospheric layer at an altitude
of 100 km. Make the approximation of a
flat Earth and calculate the time
difference between the arrivals of the
two signals.

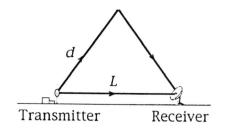

Figure 24.37

Solution: The diagonal distance from the point of transmission
to a point located directly above the midway point between
transmitter and receiver is

$$d = \sqrt{(100 \text{ km})^2 + (100 \text{ km})^2}$$

$$d = \sqrt{2}\,(100 \text{ km}) = 141.4 \text{ km}$$

The time to travel the straight line path is

$$t_1 = \frac{L}{c} = \frac{200 \times 10^3 \text{ m}}{3 \times 10^8 \text{ m / s}} = 6.67 \times 10^{-4} \text{ s}$$

and the time to travel the reflected path is

$$t_2 = \frac{2d}{c} = \frac{2(141.4 \times 10^3 \text{ m})}{3 \times 10^8 \text{ m / s}} = 9.43 \times 10^{-4} \text{ s}$$

for a time difference of $t_2 - t_1 = 0.276 \text{ ms}$ ☐

24.43 **Problem:** A dish antenna with a diameter of
20 m receives (at normal incidence) a radio
signal from a distant source, as shown in
Figure 24.43.   The radio signal is a continuous
sinusoidal wave with amplitude $E_0 = 0.2$ μV/m.
Assume the antenna absorbs all the radiation
that falls on the dish.   (a) What is the
amplitude of the magnetic field in this wave?
(b) What is the intensity of the radiation
received by this antenna?   (c) What is the power received by the
antenna? (d) What force is exerted on the antenna by the radio
waves?

Figure 24.43

Solution: (a) $B_m = \dfrac{E_m}{c} = 6.67 \times 10^{-16}$ T □

(b) $S_{av} = \dfrac{E_m^2}{2\,\mu_0 c} = 5.31 \times 10^{-17}$ W/m^2 □

(c) $P_{av} = S_{av}A = 1.67 \times 10^{-14}$ W □

Do not mix up this power with the pressure $P = S/c$, which we
use to find the force:

(d) $F = PA = \left(\dfrac{S_{av}}{c}\right)A = 5.56 \times 10^{-23}$ N □

(≈ weight of 3000 H atoms!)

24.45 **Problem:** In 1965, Penzias and Wilson discovered the cosmic microwave radiation left over from the Big Bang expansion of the Universe. The energy density of this radiation is 4×10^{-14} J/m^3. Determine the corresponding electric field amplitude.

Solution: $u = \dfrac{1}{2} \epsilon_0 E_m^2$ (Equation 24.26)

$E_m = \sqrt{\dfrac{2u}{\epsilon_0}} = 95$ mV/m $\square$

24.53 **Problem:** An astronaut, stranded in space "at rest" 10 m from his spacecraft, has a mass (including equipment) of 110 kg. He has a 100-W light source that forms a directed beam, so he decides to use the beam of light as a photon rocket to propel himself continuously toward the spacecraft. (a) Calculate how long it will take him to reach the spacecraft by this method. (b) Suppose, instead, that he decides to throw the light source away in a direction opposite to the spacecraft. If the mass of light source is 3 kg and, after being thrown, moves with a speed of 12 m/s *relative to the recoiling astronaut*, how long will the astronaut take to reach the spacecraft?

Solution:

(a) The light exerts back on him radiation pressure

$$P = \frac{F}{A} = \frac{S}{c}$$

So it exerts back on him force

$$F = \frac{SA}{c} = \frac{\text{Power}}{c} = \frac{100 \text{ J/ s}}{3 \times 10^8 \text{ m/ s}} = 3.33 \times 10^{-7} \text{ N} = (110 \text{ kg}) \ a$$

to give him acceleration $a = 3.03 \times 10^{-9} \text{ m/s}^2$.

Then he moves according to

$$X = \frac{1}{2}at^2 \qquad t = \sqrt{\frac{2x}{a}} = 8.12 \times 10^4 \text{ s} = 22.6 \text{ h} \ \square$$

(b) Momentum is conserved between an original picture of the astronaut at rest, and a final picture of a 107 kg astronaut moving at speed v and a 3 kg flashlight moving in the opposite direction at speed $v = 12$ m/s relative to the spacecraft. Thus

$0 = (107 \text{ kg})v - 3 \text{ kg}(12 \text{ m/s} - v)$

$0 = (107 \text{ kg})v - 36 \text{ kg} \cdot \text{m/s} + (3 \text{ kg})v$

$$v = \frac{36}{110} = 0.327 \text{ m/ s} \qquad t = \frac{10 \text{ m}}{0.327 \text{ m/ s}} = 30.6 \text{ s} \ \square$$

CHAPTER 25

QUESTIONS

3   **Question:** The rectangular aquarium
sketched in Figure 25.3 contains only one
gold fish. When the fish is near a corner
of the aquarium and is viewed along a
direction which makes an equal angle with
two adjacent faces, the observer sees two
fish mirroring each other, as shown.
Explain this observation.

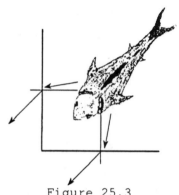

Figure 25.3

Answer: No mirrors are involved. The picture shows a top view
of the fish facing the corner of the tank. Light scattered off
the fish's right side and passing out through one wall of the
tank bends away from the normal to give you an image of the right
side of the fish, apparently displaced away from the corner.
Light from the fish's left side refracts at the other wall of the
tank to give you an image of this left side, apparently displaced
away from the corner in the opposite direction.

4   Question: As light travels from one medium to another, does its wavelength change? Does its frequency change? Does its speed change? Explain.

Answer: As light travels from one medium to another its speed changes as described by the indexes of refraction of the two media. Its frequency stays constant, since any one crest coming up to the interface produces immediately just one crest going on in the new medium. The number of cycles in a second in the first material will be the number of cycles per second in the new medium. The speed changes as does the wavelength, according to $\lambda = v/f$.

6   Question: Explain why a diamond loses most of its sparkle when submerged in carbon disulfide, and why an imitation diamond of cubic zirconia loses all of its sparkle in corn syrup.

Answer: A ball covered with mirrors sparkles by reflecting light from its surface. On the other hand, a faceted diamond lets in light at the top, reflects it by total internal reflection in the bottom half, and sends the light out through the top again. Its high index of refraction means that the critical angle for total internal reflection $\text{Arcsin}(n_{\text{air}}/n_{\text{diamond}})$ is small for a diamond in air. Thus light rays are reflected over a wide range of angles of incidence, namely all those larger than the critical angle. When the gem is immersed in carbon disulfide, the critical angle is increased to $\text{Arcsin}\left(n_{\text{CS}_2} / n_{\text{diamond}}\right)$. Then more rays coming down through the stone reach the bottom at too small an angle for total reflection, and are lost by refracting out the bottom facets. Cubic zirconia has the same index of refraction as corn syrup. This stone in this medium will reflect essentially no light, externally or internally. To a first approximation, we expect it to be invisible. To a second approximation, it may show different absorption, or a different color.

19   **Question:**   What are the conditions for the production of a
     mirage?   On a hot day, what is it that we are seeing when we
     observe "water on the road"?

     Answer: A mirage occurs when light changes direction as it moves
     between batches of air having different indices of refraction
     because of their different densities at different temperatures.
     When the sun makes a blacktop road hot, an apparent wet spot is
     due to refraction of light from the bright sky, originally headed
     a little below horizontal and then always bending up as it first
     enters and then leaves sequentially hotter, lower-density, lower-
     index layers of air closer to the road surface.

CHAPTER 25

PROBLEMS

25.5 **Problem:** An underwater scuba diver sees the Sun at an apparent angle of 45° from the vertical. Where is the Sun?

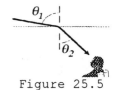

Figure 25.5

Solution: Refraction happens as sunlight in air crosses into water. The interface is horizontal so the normal is vertical

$n_1 \sin \theta_1 = n_2 \sin \theta_2$

$\sin \theta_1 = 1.33 \sin 45°$

$\sin \theta_1 = (1.33)(0.707) = 0.940$

The sunlight is at $\theta_1 = 70.5°$ to the vertical, so the Sun is 19.5° above the horizon. ☐

25.11 **Problem:** How many times will the incident beam shown in Figure 25.11 be reflected by each of the parallel mirrors?

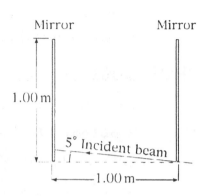

Figure 25.11

Solution: The light first hits mirror 1 (the one on the left) at distance 1 m tan 5° = 8.75 cm from its lower edge. Then the light reaches mirror 2 first at 8.75 + 8.75 = 17.5 cm from its lower edge. It returns to mirror 1 at heights of 3 x 8.75 cm, 5 x 8.75 cm, 7 x 8.75 cm,....

Since (100 cm)/(8.75 cm) = 11.4, the light last reflects from mirror 1 at 11 x 8.75 cm, which is its sixth encounter. In between, it has bounced off mirror 2 at distances 4 x 8.75, 6 x 8.75, ... 10 x 8.75 cm from the lower edge, for five reflections from this mirror. The final path of the light is to the right at 5° above the horizontal.

25.15 **Problem:** Find the time required for the light to pass through the glass block described in Problem 14.

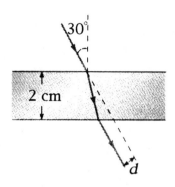

Figure 25.15

Solution: At entry the angle between ray and vertical changes to θ_2 in

$$n_1 \sin \theta_1 = n_2 \sin \theta_2$$

or $1 \sin 30 = 1.5 \sin \theta_2$

$\theta_2 = \text{Arcsin } 0.333 = 19.5°$

For the distance h traveled by the light inside the glass we have

$\cos 19.5° = (2 \text{ cm})/h.$

$$-h = \frac{2 \text{ cm}}{\cos 19.5°} = 2.12 \text{ cm}$$

The speed of light in the material is

$$v = \frac{c}{n} = \frac{3 \times 10^8 \text{ m/ s}}{1.50} = 2 \times 10^8 \text{ m/ s}$$

Therefore, $t = \frac{h}{v} = \frac{2.12 \times 10^{-2} \text{ m}}{2 \times 10^8 \text{ m/ s}} = 1.06 \times 10^{-10} \text{ s } \square$

25.25 **Problem:** A triangular glass prism with apex
angle 60° has an index of refraction of
$n = 1.5$. (a) What is the smallest angle of
incidence, θ_1, for which a light ray can
emerge from the other side? (Fig. 25.13)
(b) For what angle of incidence, θ_1, does
the light ray leave at the same angle, θ_1?

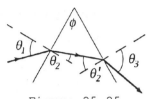

Figure 25.25

Solution: Call the angles of incidence and refraction, at the
surfaces of entry and exit, θ_1, θ_2, θ^1_2, and θ_3, in order as
shown. The apex angle $\phi = 60°$ is the angle between the
surfaces of entry and exit. The ray in the glass forms a
triangle with these surfaces, in which the interior angles must
add to 180° thus: $(90 - \theta_2) + \phi + (90 - \theta^1_2) = 180$
Then $\theta_2 + \theta^1_2 = \phi$, a general rule for light going through
prisms.

(a) The photograph on page 724 shows the effect nicely. At
the first refraction, $\sin\theta_1 = n \sin\theta_2$. Total internal
reflection occurs for

$$1.5 \sin\theta^1_2 = 1 \sin 90°$$

$$\theta^1_2 = \sin^{-1}\left(\frac{1}{1.5}\right) = \sin^{-1}(0.667) = 41.8°$$

$$\theta_2 = 60° - \theta^1_2 = 18.26°$$

Since $\sin\theta_1 = n \sin\theta_2$,

$$\theta_1 = \sin^{-1}(n \sin\theta_2) = 27.9° \ \square$$

(b) When light passes symmetrically through the prism, $\theta_1 = \theta_3$
requires $\theta_2 = \theta^1_2$. With $\theta_2 + \theta^1_2 = 60°$, we have $\theta_2 = 30°$. Then,
$\sin\theta_1 = 1.5 \sin 30$ $\theta_1 = 48.6° \ \square$. This symmetric passage
is easy to identify experimentally, since it happens to result
in minimum deviation of the light.

25.27 **Problem:** A ray of light passes from air into water. In order for its deviation angle $\delta = |\theta_1 - \theta_2|$ to be 10°, what must be its angle of incidence? You may need to use a calculator.

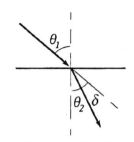

Figure 25.27

Solution: The picture suggests how $\theta_1 = \theta_2 + \delta$. Think of this as one simultaneous equation in the two unknowns θ_1 and θ_2, along with $1 \sin \theta_1 = (4/3) \sin \theta_2$. We may reduce to a single equation in one unknown with ordinary algebra, $\theta_1 = \sin^{-1}(n_2 \sin \theta_2 = \sin^{-1}[n_2 \sin(\theta_1 - 10°)]$, but this equation is transcendental: ordinary algebra cannot solve it. We can home in on the answer to any desired precision, say three digits, by directed trial and error on a hand-held calculator:

Try $\theta_1 = 40°$. Then $40° \overset{?}{=} \sin^{-1}[4/3 \sin 30°] = 41.8°$. This is greater than 40°.

Try $\theta_1 = 30°$. $30° \overset{?}{=} \sin^{-1}[4/3 \sin 20°] = 27.1°$ This is less than 30°, so the special angle must be between 30° and 40°.

Try $\theta_1 = 38°$. $38° \overset{?}{=} \sin^{-1}[4/3 \sin 28°] = 38.75°$

Try $\theta_1 = 37° \overset{?}{=} \sin^{-1}[4/3 \sin 27°] = 37.25°$

Try $\theta_1 = 36.8° = \sin^{-1}[4/3 \sin 26.8°] = 36.95°$

Try $\theta_1 = 36.5° \ \Box = \sin^{-1}[4/3 \sin 26.5°] = 36.5°$, while 36.4° and 36.6° do not work so well. So 36.5° is the answer.

25.31 **Problem:** Consider a common mirage formed by heated air just above the roadway. If an observer viewing from 2 m above the road (where $n = 1.0003$) sees water up the road at

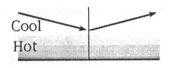

Figure 25.31

$\theta_1 = 88.8°$, find the index of refraction of the air above the road surface. (*Hint:* Treat this as a problem in total internal reflection.)

Solution: Think of the air as in two discrete layers, the first medium being cooler air with $n_1 = 1.0003$ and the second medium being hot air with a lower index, which reflects light from the sky because refraction is impossible according to

$n_1 \sin \theta_1 \geq n_2 \sin 90°$

$1.0003 \sin 88.8° \geq n_2$

$n_2 \leq 1.00008 \ \square$

25.33 **Problem:** Determine the maximum angle, θ, for which the light rays incident on the end of the pipe in Figure 25.33 are subject to total internal reflection along the walls of the pipe. Assume that the pipe has an index of refraction of 1.36 and the outside medium is air.

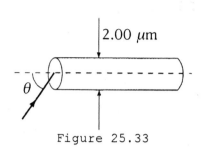

2.00 μm

Figure 25.33

Solution: At the curved wall, for total internal reflection

$$n_{pipe} \sin \theta_c = n_{air} \sin 90°$$

$$\sin \theta_c = \frac{n_{air}}{n_{pipe}} = \frac{1.00}{1.36} = 0.7353$$

$$\theta_c = 47.3°$$

Geometry shows that the angle of refraction at the end is

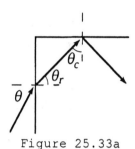

Figure 25.33a

$$\theta_r = 90° - \theta_c = 90° - 47.3° = 42.7°$$

Then, Snell's law at the end,

1.00 $\sin \theta$ = 1.36 $\sin 42.7°$, gives $\theta = 67.2°$ □

25.37 **Problem:** A small underwater pool light is 1 m below the

surface. What is the maximum possible value

for the angle of refraction?

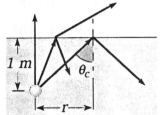

Solution: At the edge of the circle, the

light is totally internally reflected:

$n_1 \sin \theta_c = n_2 \sin 90°$

Figure 25.37

$(4/3) \sin \theta_c = 1$

$\sin\theta_c = \dfrac{1}{4 \, / \, 3} = \dfrac{3}{4}$

$\theta_c = \sin^{-1}(0.750) = 48.6°$

The radius then satisfies $\tan \theta_c = r/(1 \text{ m})$. So the diameter is

$d = 2 \tan \theta_c = 2 \tan 48.6° = 2.27 \text{ m}$ □

25.41 **Problem:** The laws of refraction and reflection are the same for sound as for light. The speed of sound in air is 340 m/s, and that in water is 1510 m/s. If a sound wave approaches a plane water surface at an angle of incidence of 12°, what is the angle of refraction?

Solution: With $v = \dfrac{c}{n}$, we put Snell's law into a form referring to speeds:

$n_1 \sin\theta_1 = n_2 \sin\theta_2$

$\dfrac{c}{v_1}\sin\theta_1 = \dfrac{c}{v_2}\sin\theta_2.$

$\dfrac{\sin\theta}{v_1} = \dfrac{\sin\theta_2}{v_2}$

This is true also for sound. Here

$\dfrac{\sin 12°}{340 \text{ m/s}} = \dfrac{\sin\theta_2}{1510 \text{ m/s}}$

$\theta_2 = (\text{arcsin } 4.44°)\sin 12°$

$\theta_2 = 67.4° \;\square$

25.47   **Problem:** A hiker stands on a mountain peak near sunset and observes a (primary) rainbow caused by water droplets in the air about 8 km away. The valley is 2 km

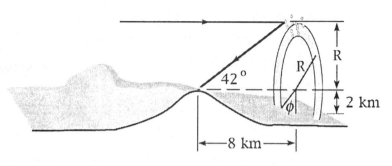

Figure 25.47

below the mountain peak and is entirely flat. What fraction of the complete circular arc of the rainbow is visible to the hiker? (See Question 16.)

Solution: Horizontal light rays from the setting sun pass the hiker at P, are twice refracted and once reflected as in the text's figure 25.25 by just the certain special raindrops at 40° to 42° from the hiker's shadow, and reach the hiker as the rainbow. The hiker sees more of the violet inner edge, so we consider the red outer edge. The radius R of the circle of droplets is R = 8 km sin 42° = 5.35 km. Then the angle ϕ, between the vertical and the radius where the bow touches the ground, is given by cos ϕ = (2 km)$/R$ = (2 km)/(5.35 km) = 0.374 as ϕ = 68.1°. The angle filled by the visible bow is 360° - (2 x 68.1°) = 223.8°, so the visible bow is

$$\frac{223.8°}{360°} = 62.2\% \text{ of a circle } \square$$

This striking view motivated Charles Wilson's 1906 invention of the cloud chamber, a standard tool of nuclear physics. Look for a full-circle rainbow around your shadow when you fly in an airplane.

25.51 **Problem:** A laser beam strikes one end of a slab of material, as shown in Figure 25.51. The index of refraction of the slab

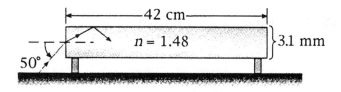

Figure 25.51

is 1.48. Determine the number of internal reflections of the beam before it emerges from the opposite end of the slab.

Solution: On entrance,

$\sin 50° = 1.48 \sin \theta_2$

$\theta_2 = 31.17°$

The beam strikes the top face at horizontal coordinate

$x_1 = \dfrac{1.55 \text{ mm}}{\tan 31.17°} = 2.562 \text{ mm}$

Thereafter, the beam strikes a face every $2x_1 = 5.124$ mm.

Since the slab is 420 mm long, the beam makes $\dfrac{420 - 2.562}{5.124} = 81$

more reflections, for a total of 82 reflections. ☐

25.53   **Problem:**   The light beam in figure 25.53
strikes surface 2 at the critical angle.
Determine the angle of incidence, θ_1.

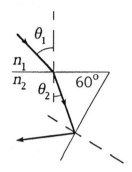

Solution: Call the index of refraction of the
prism material n_2.   Use n_1 to represent the
surrounding stuff.   At surface 2,

$n_2 \sin 42° = n_1 \sin 90°$

Figure 25.53

$$n_2/n_1 = \frac{1}{\sin 42°} = 1.5557$$

Call the angle of refraction θ_2 at the surface 1.   The ray
inside the prism forms with surfaces 1 and 2 a triangle whose
interior angles add to 180° thus:

$(90 - \theta_2) + 60 + (90 - 342) = 180$

$60 = \theta_2 + 42$

$\theta_2 = 18°$

At surface 1,

$n_1 \sin \theta_1 = n_2 \sin 18°$

$\sin \theta_1 = (n_2/n_1) \sin 18°$

$$\sin \theta_1 = \frac{\sin 18°}{\sin 42°}$$

$\theta_1 = 28.7°$ □

CHAPTER 26

QUESTIONS

2   **Question:** The side-view mirror on late-model cars warns the user that objects may be closer than they appear. What kind of mirror is being used, and why was that type selected?

Answer: The mirror is convex to give a wide-angle view.

11   **Question:** Describe lenses that can be used to start a fire.

Answer: The burning glass must make a real image of the distant Sun, making nearly parallel rays converge. Any converging lens will do. A large-diameter lens will gather more power to start the fire faster.

13   **Question:** Consider the image formed by a thin converging lens. Under what conditions will the image be (a) inverted? (b) erect? (c) real? (d) virtual? (e) larger than the object? (f) smaller than the object?

Answer: For definiteness, we consider real objects ($p > 0$).

(a)   For $M = -q/p$ to be negative, q must be positive. In $1/q = 1/f - 1/p$, this will happen if $p > f$: if the object is further than the focal point.

(b)   For $M = -q/p$ to be positive; q must be negative. From $1/q = 1/f - 1/p$. we need $p < f$.

(c)   For a real image, q must be positive. As in part (a), it is sufficient for p to be larger than f.

(d)   For $q < 0$ we need $p < f$.

(e)   For $|M| > 1$, we consider separately $M < -1$ and $M > 1$. If $M = -q/p < -1$, we need $q/p > 1$; $q > p$; $1/q < 1/p$. From $1/p + 1/q = 1/f$, we get $1/p + 1/p > 1/f$; $2/p > 1/f$; $p/2 < f$; $p < 2f$. Now if $-q/p > 1$; $-q > p$; $q < -p$ we may require $q < 0$, since then $1/p = 1/f - 1/q$ with $1/f > 0$ gives $1/p > -1q$; $-p > q$ as required. For $q < 0$ in $1/q = 1/f - 1/p$, we need $p < f$. Thus the overall condition for an enlarged image is simply $p < 2f$.

(f)   For $|M| < 1$, we have the reverse of part (e), requiring $p > 2f$.

19   **Question:** A solar furnace can be constructed by using a concave mirror to reflect and focus sunlight into a furnace enclosure. What factors in the design of the reflecting mirror will guarantee that very high temperatures can be achieved?

Answer: Make the mirror an efficient reflector (shiny); make it reflect to the image even rays far from the axis (by giving it a parabolic shape); most important, make it large in diameter to intercept a lot of solar power. And you get a higher temperature if the image is smaller, as you get with shorter focal length; and if the furnace is an efficient absorber (black).

CHAPTER 26

PROBLEMS

26.3   Problem: Determine the minimum height of a vertical plane mirror in which a person 5'10" in height can see his or her full image. (A ray diagram would be helpful.)

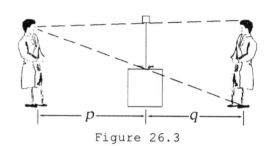

Figure 26.3

Solution: The flatness of the mirror is described by $R = \infty$, $f = \infty$, $1/f = 0$. Then in $1/p + 1/q = 1/f$, we have $q = -p$: the image is as far behind the mirror as you are in front. The magnification is then $-q/p = 1 = h'/h$ so $h' = h = 5'10" = 70"$. The height of mirror required is the part through which you look (at distance p) to see a 70" high object at distance $2p$. The mirror height is $(\frac{1}{2})70" = 35" = 2'11"$.

26.9   Problem: A concave mirror has a focal length of 40 cm. Determine the object position for which the resulting image will be erect and four times the size of the object.

Solution: For a concave mirror, R and f are positive. Also, for an erect image, M is positive. Therefore, $M = \dfrac{-q}{p} = 4$ and $q = -4p$. The image is virtual.

$$\frac{1}{f} = \frac{1}{p} + \frac{1}{q} \quad \text{becomes} \quad \frac{1}{40 \text{ cm}} = \frac{1}{p} - \frac{1}{4p} = \frac{3}{4p}$$

From which, $p = 30$ cm $\square$

26.13 **Problem:** A spherical mirror is to be used to form an image five times the size of an object on a screen positioned 5 m from the object. (a) Describe the type of mirror required. (b) Where should the mirror be positioned relative to the object?

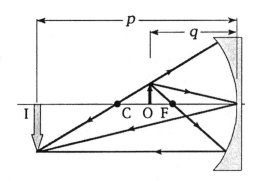

Figure 26.13

Solution: $q = (p + 5 \text{ m})$ and, since the image must be real,

$M = -\dfrac{q}{p} = -5$ or $q = 5p$. Therefore, $p + 5 = 5p$ or $p = 1.25 \text{ m}$.

This is the answer to part b: the mirror should be 1.25 m from the object □. Now $q = 5p = 6.25 \text{ m}$. The image is real, inverted, and enlarged. The mirror must have focal length given by $1/f = 1/p + 1/q = 1/(1.25 \text{ m}) + 1/(6.25 \text{ m})$
$f = 1.04 \text{ m}$. It must be a concave mirror of radius
$R = 2f = 2.08 \text{ m}$ □

26.23 Problem: A glass hemisphere is used as
a paperweight, with its flat face
resting on a stack of papers. The
radius of the circular cross-section is
4 cm, and the index of refraction of the
glass is 1.55. The center of the hemisphere is directly over a letter "O" that
is 2.5 mm in diameter. What is the dia-
meter of the image of the letter as seen from above along a
vertical radius?

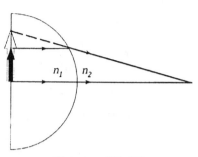

Figure 26.23

Solution: The image forms by refraction at a single surface:

$n_1/p + n_2/q = (n_2 - n_1)/R$

$1.55/(4 \text{ cm}) + 1/q = (1 - 1.55)/(-4 \text{ cm})$

$q = 1/(0.55/4 \text{ cm} - 1.55/4 \text{ cm}) = -4 \text{ cm}$

The virtual image is at the same location as the object.

Magnification $M = - \dfrac{n_1 q}{n_2 p} = - \dfrac{(1.55)(-4 \text{ cm})}{(1)(4 \text{ cm})} = 1.55$

and the image height, $h' = Mh = 1.55(2.5 \text{ mm}) = 3.88 \text{ mm}$ □

The image is virtual, erect, and enlarged.

26.27 **Problem:** The left face of a biconvex lens has a radius of curvature of 12 cm, and the right face has a radius of curvature of 18 cm. The index of refraction of the glass is 1.44. (a) Calculate the focal length of the lens. (b) Calculate the focal length if the radii of curvature of the two faces are interchanged.

Solution: Convex outward on both sides, the lens has the centers of curvature of its surfaces on opposite sides. The second surface has negative radius:

(a) $\dfrac{1}{f} = (n - 1)\left[\dfrac{1}{R_1} - \dfrac{1}{R_2}\right] = (1.44 - 1)\left[\dfrac{1}{(12\ \text{cm})} - \dfrac{1}{(-18\ \text{cm})}\right]$

and $f = 16.4$ □

(b) $\dfrac{1}{f} = (0.44)\left[\dfrac{1}{(18\ \text{cm})} - \dfrac{1}{(-12\ \text{cm})}\right]$ and $f = 16.4$ cm □

The lens can be reversed without changing what it does to the light! In laboratory take any asymmetric biconvex lens and prove this to yourself, your lab partner, and your instructor.

26.33 **Problem:** A microscope
slide is placed in front
of a converging lens
with a focal length of
2.44 cm.   The lens forms
an image of the slide
12.9 cm from the slide.
How far is the lens from
the slide if the image is (a) real?   (b) virtual?

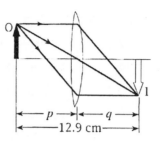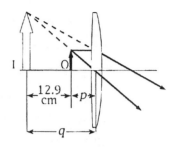

Figure 26.33

Solution:

(a)   The lens is between slide and image, so

$p + q = 12.9$ cm. Thus $q = 12.9$ cm $- p$

so $\dfrac{1}{p} + \dfrac{1}{12.9 - p} = \dfrac{1}{2.44}$

To solve, simplify:

$\dfrac{12.9 - p + p}{12.9p - p^2} = \dfrac{1}{2.44}$

$31.48 = 12.9p - p^2$

$p^2 - 12.9p + 31.48 = 0$

$p = \dfrac{12.9 \pm \sqrt{12.9^2 - 4 \cdot 31.48}}{2}$

$p = 9.63$ cm or $p = 3.27$ cm $\square$

Both solutions are valid

(b) $|q| = -q = p + 12.9$ cm

$\dfrac{1}{p} - \dfrac{1}{12.9 + p} = \dfrac{1}{2.44}$ or

$p^2 + 12.9p = 31.8$

from which $p = 2.1$ cm or $p = -15$ cm.   We must have
a real object so the -15 cm solution must be
rejected.

26.39 **Problem:** A person looks at a gem with a jeweler's microscope --a converging lens with a focal length of 12.5 cm. The microscope forms a virtual image 30.0 cm from the lens. Determine the magnification of the lens. Is the image upright or inverted?

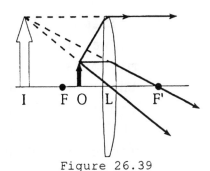

Figure 26.39

Solution: $\dfrac{1}{p} + \dfrac{1}{q} = \dfrac{1}{f}$

$$\frac{1}{p} + \frac{1}{-30} = \frac{1}{12.5}$$

$$p = 8.8235 \text{ cm}$$

$$M = \frac{-q}{p} = -\frac{-30}{8.82} = 3.40, \text{ upright } \square$$

26.43 **Problem:** The object in Figure 26.43 is midway between the lens and the mirror. The mirror's radius of curvature is 20.0 cm, and the lens has a focal length of -16.7 cm. Considering only the light that leaves the object and travels first toward the mirror, locate the final image formed by this system. Is this image real or virtual? Is it erect or inverted? What is the overall magnification?

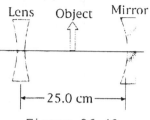

Figure 26.43

Solution: The mirror has $f = R/2 = +10$ cm and forms an image thus:

$$\frac{1}{q_1} = \frac{1}{f_1} - \frac{1}{p_1} = \frac{1}{10 \text{ cm}} - \frac{1}{12.5 \text{ cm}}$$

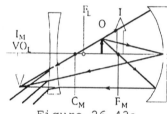

Figure 26.43a

$q_1 = 50$ cm (to left of mirror)

The light rays heading toward this real image are intercepted before they make it by the lens, which sees them as object rays converging to a virtual object behind the lens by 50 - 25 = 25 cm, at $p = -25$ cm. Then the lens forms the final image:

$$\frac{1}{q_2} = \frac{1}{f_2} - \frac{1}{p_2} = \frac{1}{-16.7 \text{ cm}} - \frac{1}{-25 \text{ cm}} q_2$$

$$\frac{1}{q_2} = -50.3 \text{ cm (to right of lens)}$$

Thus, the final image is located 25.3 cm to right of mirror ☐

$$M_1 = -\frac{q_1}{p_1} = -\frac{50 \text{ cm}}{12.5 \text{ cm}} = -4$$

$$M_2 = -\frac{q_2}{p_2} = -\frac{-50.3 \text{ cm}}{-25 \text{ cm}} = -2.01$$

The overall magnification is $M = M_1 M_2 = 8.05$ ☐

Thus, the final image is virtual, erect, and 8.05 times the size of object, and 25.3 cm to right of the mirror.

26.47 **Problem:** A philatelist examines the
printing detail on a stamp using a
convex lens of focal length 10 cm as a
simple magnifier. The lens is held
close to the eye, and the lens-to-object
distance is adjusted so that the virtual
image is formed at the normal near
point, 25 cm from the eye. Calculate
the expected magnification.

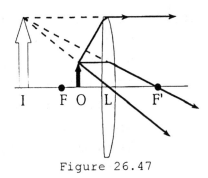

Figure 26.47

Solution: The image is virtual: $q = -25$ cm
The distance from stamp to magnifying glass is p in

$$\frac{1}{p} + \frac{1}{q} = \frac{1}{f} \qquad \frac{1}{p} + \frac{1}{-25 \text{ cm}} = \frac{1}{10 \text{ cm}}$$

$p = 7.14$ cm

Then

$M = -q/p = -(-25 \text{ cm})/(7.14 \text{ cm})$

$M = 3.50$ □

26.51 **Problem:** A parallel beam of
light enters a glass hemisphere
perpendicular to the flat face,
as shown in Figure 26.51. The
radius is $R = 6$ cm, and the index
of refraction is $n = 1.560$.
Determine the point at which the
beam is focused. (Assume paraxial rays.)

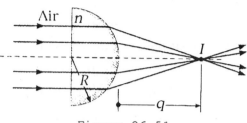

Figure 26.51

Solution: The hemisphere is too thick to be described as a
thin lens. The light is undeviated on entry into the flat
face. We need only consider its exit from the second surface,
for which $R = -6$ cm. The incident rays are parallel, as
described by

$p = \infty$. Then $n_1/p + n_2/q = (n_2 - n_1)/R$ becomes

$0 + 1/q = (1 - 1.56)/(-6 \text{ cm})$

$q = 10.7$ cm $\square$

26.55 **Problem:** The disk of the Sun subtends an angle of 0.5 degrees at the Earth. What are the position and diameter of the solar image formed by a concave spherical mirror of radius 3 m?

Solution: For the mirror $f = R/2 = +1.5$ m. In $\frac{1}{p} + \frac{1}{q} = \frac{1}{f}$ the distance to the Sun is so much larger that we can take $p = \infty$:

$\frac{1}{\infty} + \frac{1}{q} = \frac{1}{f}$ to locate the image as $q = 1.5$ m $\square$, from the mirror.

Now in $M = -\frac{q}{p} = \frac{h'}{h}$ the magnification is nearly zero, but we can be more precise: h/p is the angular diameter of the object. So the image diameter is

$$h' = -\frac{hq}{p} = -0.5° \left(\frac{\pi \text{radian}}{180°} \right) 1.5 \text{ m} = -1.31 \text{ cm } \square$$

26.59 **Problem:** In a darkened room, a burning
candle is placed 1.5 m from a white
wall. A lens is placed between the
candle and wall at a location that
causes a larger, inverted image of the
candle to form on the wall. When the
lens is moved 90 cm toward the wall,
another image of the candle is formed.

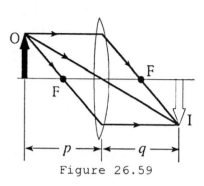

Figure 26.59

Find (a) the two object distances that produce the images just
described and (b) the focal length of the lens. (c) Charac-
terize the second image.

Solution: In the original situation, $p_1 + q_1 = 1.5$ m. In the
final situation, $p_2 = p_1 + 0.9$ m,
$q_2 = q_1 - 0.9$ m.

And $\dfrac{1}{p_1} + \dfrac{1}{q_1} = \dfrac{1}{f} = \dfrac{1}{p_2} + \dfrac{1}{q_2}$

Thus, substitution gives

$$\frac{1}{p_1} + \frac{1}{1.5 - p_1} = \frac{1}{p_1 + 0.9} + \frac{1}{0.6 - p_1}$$

$$\frac{1.5 - p_1 + p_1}{1.5\,p_1 - p_1^2} = \frac{0.6 - p_1 + p_1 + 0.9}{0.6\,p_1 - p_1^2 + 0.54 - 0.9\,p_1}$$

$$1.5p_1 - p_1{}^2 = -0.3p_1 - p_1{}^2 + 0.54$$

$p_1 = 0.54/1.8 = 0.300$ m □ and

$p_2 = p_1 + 0.9 = 1.20$ m □

(b) $1/f = 1/(0.3 \text{ m}) + 1/(1.5 - 0.3)\text{m}$

        $f = 0.240$ m □

(c)   The second image is real, inverted, and diminished □
        with $M = -q_2/p_2 = -0.250$

CHAPTER 27

QUESTIONS

4   **Question:** Why is it so much easier to perform interference experiments with a laser than with an ordinary light source?

Answer: You first pass light from an ordinary source through a prism or diffraction grating (described in the next chapter) to disperse different colors into different directions. With a single narrow slit you select a single color and make that light diffract (spread -- see the next chapter) to cover both of the slits for Young's experiment. Thus you may have trouble lining things up and you will generally have low light power reaching the screen. The laser light is already monochromatic and coherent across the width of the beam.

5   **Question:** In Young's double-slit experiment, why do we use monochromatic light? If white light were used, how would the pattern change?

Answer: Every color produces its own pattern and we see them superimposed. The central maximum is white. The first side maximum is a full spectrum, with violet on the inside and red on the outside. The second side maximum is a full spectrum also, but red in it overlaps with violet in the third maximum. At larger angles the light soon starts mixing to white again -- or it is so faint you might say gray.

13   **Question:** Would it be possible to place a nonreflective
        coating on an airplane to cancel radar waves of wavelength
        3 cm?

        Answer: Make an antireflective coating for radar like this:
        Measure the radar-reflectivity of the metal of your airplane.
        Suppose it is 90%. Then choose a light durable material that
        will reflect just about 45% of the radio-wave energy incident
        on it. Measure its index of refraction and onto the metal
        plaster a coating equal in thickness to one-quarter of 3 cm
        divided by that index. Sell it quick and then you can sell to
        the supposed enemy new radars operating at 1.5 cm, which the
        coated metal will reflect with extra-high efficiency.

16   **Question:** Suppose we use reflected white light to observe a
        thin, transparent coating on glass as the coating material is
        gradually deposited by evaporation in a vacuum. Describe
        possible color changes that might occur during the process of
        building up the thickness of the coating.

        Answer: Suppose the coating is intermediate in index of
        refraction between vacuum and the glass. When the coating is
        very thin, light reflected from its top and bottom surfaces
        will interfere constructively, so you see the surface white and
        brighter. As the thickness reaches one-quarter of the
        wavelength of violet light in the coating, destructive
        interference for violet will make the surface look red. Next
        to interfere destructively are blue, green, yellow, orange, and
        red, making the surface look red, purple and then blue. As the
        coating gets still thicker, we can get constructive
        interference for violet and then for other colors in spectral
        order. Still thicker coatings will give constructive and
        destructive interference for several visible wavelengths, so
        the reflected light starts looking white again.

CHAPTER 27

PROBLEMS

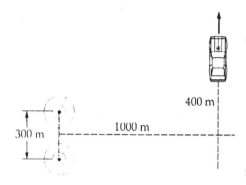

27.5 **Problem:** Two radio antennas separated by 300 m, as shown in Figure 27.5, simultaneously transmit identical signals (assume waves) on the same wavelength. A radio in a car traveling due north receives the signals. (a) If the car is at the position of the second maximum, what is the wave-length of the signals? (b) How much farther must the car travel to encounter the next minimum in reception? (*Caution:* Avoid small-angle approximations in this problem.)

Figure 27.5

Solution: (a) The distance of the car from the two antennas is

$$d_1 = \sqrt{(1000)^2 + (250)^2} = 1030.776 \text{ m}$$

$$d_2 = \sqrt{(1000)^2 + (550)^2} = 1141.271 \text{ m}$$

Being at the second side maximum means

$$\delta = d_2 - d_1 = 2\lambda \qquad \lambda = 55.25 \text{ m} \quad \square$$

(b) Call y the extra distance. At the next minimum we have

$$2.5\lambda = \sqrt{(1000)^2 + (400 + y + 150)^2} - \sqrt{(1000)^2 + (400 + y - 150)^2}$$

$$138.119 = \sqrt{(1000)^2 + (550 + y)^2} - \sqrt{(1000)^2 + (250 + y)^2}$$

27.5 (cont)

One can solve interatively using a small computer or
square both sides of

$$38.119 + \sqrt{1,062,500 + 500y + y^2} = \sqrt{1,302,500 + 1100y + y^2}$$

$$19076.725 + 1,062,500 + 500y + 1100y + y^2$$

$$+ 276.137404 \sqrt{1,062,500 + 500y + y^2}$$

$$= 1,302,500 + 1100y + y^2$$

$$276.23704\sqrt{1,062,500 + 500y + y^2} = 220,923.27 + 600y$$

$$\sqrt{1,062,500 + 500y + y^2} = 799.75978 + 2.1720476y$$

$$1,062,500 + 500y + y^2$$

$$= 639615.71 + 3474.2326y + 4.7177908y^2$$

$$3.7177908y^2 + 2974.2326y - 422884.29 = 0$$

$$y^2 + 800y - 113746.12 = 0$$

$$y = \frac{-800 \pm \sqrt{800^2 + 4 \times 113746.12}}{2}$$

$$y = \frac{-800 \pm 1046.4151}{2} = 123.2 \text{ m } \square \text{ or } -923.2 \text{ m}$$

The negative root represents the symmetrically located minimum
below the midline in the picture. The problem warns us not to
use the approximation $L \gg 300$ m, for 300 m $\sin \theta = m\lambda$; but if
we are satisfied with two-digit precision it is good enough,
giving $(300 \text{ m}) \sin (\text{Arctan } 0.4) = 2\lambda$; $\lambda = 55.7$ m;
$(300 \text{ m}) \sin (\text{Arctan } [(400 + y)/1000]) = 2.5(55.7 \text{ m})$;
$y = 124$ m. But note that $\sin \theta \cong \tan \theta$ does not work well at
all for the large angles here.

27.9 **Problem:** An oscillator drives two loudspeakers 35 cm apart, which vibrate in phase at a frequency of 2 kHz. At what angles, measured from the perpendicular bisector of the line joining the speakers, would a distant observer hear maximum sound intensity? Minimum? (Take the speed of sound as 340 m/s.)

Solution: $\lambda = \dfrac{340 \text{ m/s}}{2000 \text{ Hz}} = 0.17 \text{ m}$

Maxima at $d \sin \theta = m\lambda$

$m = 0$ gives $\theta = 0$

$m = 1$ gives $\sin \theta = \dfrac{\lambda}{d} = \dfrac{0.17 \text{ m}}{0.35 \text{ m}}$ $\theta = 29.1°$

$m = 2$ gives $\sin \theta = \dfrac{2\lambda}{d} = 0.971$ $\theta = 76.2°$

$m = 3$ gives $\sin \theta = 1.46$ No Solution

Minima at $d \sin \theta = \left(m + \dfrac{1}{2} \right)\lambda$

$m = 0$ gives $\sin \theta = \dfrac{\lambda}{2d} = 0.234$ $\theta = 14.1°$

$m = 1$ gives $\sin \theta = \dfrac{3\lambda}{2d} = 0.729$ $\theta = 46.8°$

$m = 2$ gives no solution.

So we have maxima at 0°, 29.1°, and 76.2° and minima at 14.1° and 46.8°. ☐

27.11 Problem: In the arrangement of Figure 27.11, let $L = 120$ cm and $d = 0.25$ cm. The slits are illuminated with light of wave-length 600 nm. Calculate the distance y above the central maximum for which the average intensity on the screen will be 75% of the maximum.

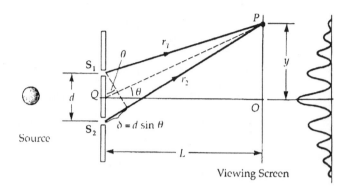

Figure 27.11

Solution:

$$I_{av} = I_0 \cos^2\left(\frac{\pi d \sin\theta}{\lambda}\right) \quad \text{for small } \theta$$

$$\sin\theta = \frac{y}{L} \quad \text{and} \quad I_{av} = 0.75 I_0$$

$$y = \frac{\lambda L}{\pi d} \cos^{-1}\left(\frac{I_{av}}{I_0}\right)^{1/2}$$

$$y = \frac{(6.0 \times 10^{-7})(1.2 \text{ m})}{\pi(2.5 \times 10^{-3} \text{ m})} \cos^{-1}\left(\frac{0.75 I_0}{I_0}\right)^{1/2}$$

$$y = 4.80 \times 10^{-5} \text{ m } \square$$

27.17 Problem: Coherent light from a helium-neon laser ($\lambda = 632.8$ nm) is incident on two parallel slits 0.2 mm apart. What is the distance to the first maximum and its intensity (relative to the central maximum) on a screen 2 m beyond the slits?

Solution: Use $d \sin q = m\lambda$ (Equation 27.2) for small angles (with $\sin\theta \cong y/L$) and for the first side maximum ($m = 1$) to obtain

$$y_{bright} = \frac{\lambda L}{d} = \frac{(632.8 \times 10^{-9})(2)}{2 \times 10^{-4}} = 6.328 \times 10^{-3} \text{ m}$$

$y_{bright} = \pm 6.33$ mm $\square$. For small θ, equal intensity at center of each maximum.

27.21 **Problem:** Determine the resultant of the two waves $E_1 = 6 \sin (100 \pi t)$ and $E_2 = 8 \sin (100 \pi t + \pi/2)$.

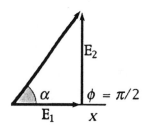

Figure 27.21

Solution: Let the x-axis lie along $\mathbf{E}_1$ in phase space. Then its component form is $\mathbf{E}_1 = 6\mathbf{i} + 0\mathbf{j}$. The components of $\mathbf{E}_2$ are

$\mathbf{E}_2 = 8 \cos \pi/2 \ \mathbf{i} + 8 \sin \pi/2 \ \mathbf{j} = 0 + 8\mathbf{j}$. The resultant is then

$\mathbf{E}_R = 6\mathbf{i} + 8\mathbf{j}$, with an amplitude of $\sqrt{6^2 + 8^2} = 10$, a phase from $\tan \alpha = 8/6$, $\alpha = 0.927$ rad, and a functional representation

$E_R = 10 \sin (100\pi t + 0.927)$ $\square$

27.29 **Problem:** A thin layer of liquid methylene iodine ($n = 1.756$) is sandwiched between two flat parallel plates of glass. What must be the thickness of the liquid layer if normally incident light with $\lambda = 600$ nm is to be strongly reflected?

Solution: A phase change of π occurs upon reflection at the first interface but not at the second. The total phase shift of the second reflected wave relative to the first is then

$$\phi = 2\pi = \frac{2nt}{\lambda} (2\pi) + \pi$$

$$t = \frac{\lambda}{4n} = \frac{(600 \times 10^{-9} \ \text{m})}{4(1.756)} = 85.4 \times 10^{-9} \ \text{m} \ \square$$

27.31   **Problem:**   An oil film (n = 1.45)
floating on water is illuminated by
white light at normal incidence.   The
film is 280 nm thick. Find (a) the
dominant observed color in the reflected
light and (b) the dominant color in the
transmitted light.   Explain your
reasoning.

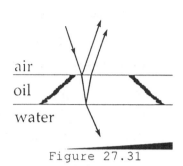

Figure 27.31

Solution: The light reflected from the top of the oil film
undergoes phase reversal.   Since 1.45 > 1.33, the light
reflected from the bottom undergoes no reversal.   For
constructive interference of reflected light we then have

$$2nt = \left(m + \frac{1}{2}\right)\lambda$$

$$\lambda = \frac{2nt}{\left(m + \frac{1}{2}\right)} = \frac{(560 \text{ nm})}{\left(m + \frac{1}{2}\right)}(1.45)$$

for $m = 0$, $\lambda = 1624$ nm   (infrared)

$m = 1$, $\lambda = 541$ nm   (yellow)

$m = 2$, $\lambda = 325$ nm   (ultraviolet)

So the dominant reflected color is yellow □

(b)   Subtracting yellow from white gives red + blue + violet =
purple □.   We could also note that the reflected light
contains little red and violet according to the condition for
destructive interference $2nt = m\lambda$

$m = 1$, $\lambda = 812$ nm   (near infrared)

$m = 2$, $\lambda = 406$ nm   (violet)

So red and violet make it through to be in the transmitted
beam.

27.39 **Problem:** In an application of interference effects to radio-astronomy,
Australian astronomers observed a 60-MHz
radio source both directly from the source
and from its reflection from the sea. If
the receiving dish is 20 m above sea
level, what is the angle of the radio
source above the horizon at first maximum?

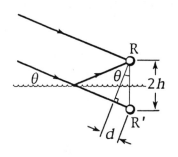

Figure 27.39

Solution: One radio wave reaches the receiver R directly from
the distant source at angle θ above the horizontal. The other
wave reflects from the water at P, there undergoing phase
reversal. Constructive interference first occurs for a path
difference $d = \lambda/2$. The two angles θ in the picture are equal
because their sides are perpendicular right side to right side
and left side to left side. It is equally far from P to R as
from P to R', the mirror image of the telescope. So the path
difference is $d = 2h \sin \theta$ where $h = 20$ m. The wavelength is

$$\lambda = \frac{c}{f} = \frac{3 \times 10^8}{6 \times 10^7} = 5 \text{ m and } 2h \sin \theta = \lambda/2 \text{ gives}$$

$$\sin \theta = \frac{\lambda}{4h} = \frac{5 \text{ m}}{80 \text{ m}} \qquad \theta = 3.58° \; \square$$

27.41 **Problem:** Measurements are made
       of the intensity distribution
       in a Young's interference
       pattern (as illustrated in
       Figure 27.41). At a particular
       value of y (the distance from

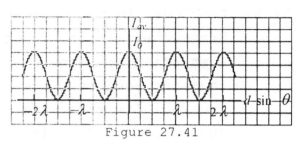

Figure 27.41

the center of the screen), it is found that $I/I_0 = 0.81$ when
light of wave-length 600 nm is used. What wavelength of light
should be used to reduce the relative intensity at the same
location to 64%?

Solution: From Equation 27.13, $\dfrac{I_1}{I_0} = \cos^2\left(\dfrac{\pi y d}{\lambda_1 L}\right) = 0.81$

Then $\dfrac{\pi y d}{L} = \lambda_1 \cos^{-1}\left(\dfrac{I_1}{I_0}\right)^{\frac{1}{2}} = (600 \text{ nm}) \cos^{-1}(0.9) = 270.6 \text{ nm}$

Let λ_2 equal the wavelength for which

$$\frac{I_1}{I_0} \rightarrow \frac{I_2}{I_0} = 0.64 = \cos^2\left(\frac{\pi y d}{\lambda_2 L}\right)$$

then $\lambda_2 = \dfrac{\pi y d \;/\; L}{\cos^{-1}(I_2 \,/\, I_0)^{1/2}}$

Substituting, we find

$$\lambda_2 = \frac{270.6 \text{ nm}}{\cos^{-1}(0.64)^{1/2}} = 421 \text{ nm} \quad \square$$

27.43 **Problem:** Young's double-slit experiment is performed with sodium yellow light (λ = 5890Å) and with a slits-to-screen distance of 2.0 m. The tenth interference minimum (dark fringe) is observed to be 7.26 mm from the central maximum. Determine the spacing of the slits.

Solution: In the equation $d \sin \theta = (m + \frac{1}{2})\lambda$, the first minimum is described by $m = 0$ and the tenth by $m = 9$. So

$$\sin \theta = \frac{\lambda}{d}\left(9 + \frac{1}{2}\right)$$

Also, $\tan \theta = \dfrac{y}{L}$

For small θ, $\sin \theta = \tan \theta$. Thus,

$$d = \frac{9.5\lambda}{\sin \theta} = \frac{9.5\lambda L}{y} = \frac{9.5(5890 \times 10^{-10} \text{ m})(2.0 \text{ m})}{7.26 \times 10^{-3} \text{ m}} = 1.54 \times 10^{-3} \text{ m}$$

$d = 1.54$ mm $\square$

27.45 **Problem:** A hair is placed at one edge between two flat glass plates 8 cm long. When this arrangement is illuminated with yellow light of wavelength 600 nm, a total of 121 dark bands are counted, starting at the point of contact of the two plates. How thick is the hair?

Solution: Light reflecting from the bottom surface of the top plate undergoes no phase shift, while light reflecting from the top surface of the bottom plate is phase shifted by π upon reflection and also has to travel extra distance $2t$, where t is the thickness of the air wedge. For destructive interference,

$2t = m\lambda$ ($m = 0$, 1, 2, 3, ...).

Therefore, for 121 dark bands, $2t = 120\lambda$

$t = 60\lambda = 3.6 \times 10^{-5}$ m $\square$

27.55 **Problem:** Consider the double-slit arrangement shown in Figure 27.55, where the separation of the slits, d, is 0.30 mm and the distance L to the screen is 1 m. A thin sheet of transparent plastic, of thickness 0.050 mm (about the thickness of this page)

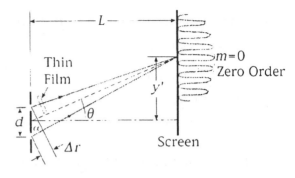

Figure 27.55

and refractive index $n = 1.50$, is placed over only the upper slit. As a result, the central maximum of the interference pattern moves upward a distance of y'. Find this distance.

Solution: Call b the thickness of the film. The central maximum corresponds to zero phase difference. Thus, the added distance Δr traveled by the light from the lower slit must introduce a phase difference equal to that introduced by the plastic film. The *phase difference* ϕ is

$$\phi = 2\pi\left(\frac{b}{\lambda_a}\right)(n - 1)$$

The corresponding difference in *path length* Δr is

$$\Delta r = \phi\left(\frac{\lambda_a}{2\pi}\right) = 2\pi\left(\frac{b}{\lambda_a}\right)(n - 1)\left(\frac{\lambda_a}{2\pi}\right) = b(n - 1)$$

Note that the wavelength of the light does not appear in this equation. In Figure 27.55, the two rays from the slits are essentially parallel, so the angle θ may be expressed as

$$\tan\theta = \frac{\Delta r}{d} = \frac{y'}{L}$$

Eliminating Δr by substitution,

$y'/L = b(n - 1)/d$ gives

$y' = b(n - 1)L/d$

$$y' = \frac{(5 \times 10^{-5} \text{ m})(1.50 - 1)(1 \text{ m})}{(3 \times 10^{-4} \text{ m})}$$

$y' = 0.0833$ m $= 8.33$ cm $\square$

CHAPTER 28

QUESTIONS

3   **Question:**   Although we can hear around corners, we cannot see around coners.   How can you explain this in view of the fact that sound and light are both waves?

Answer: Audible sound has wavelengths on the order of meters or centimeters, while visible light has wavelength on the order of half a micrometer.   In this world of breadbox-size objects, λ/a is large for sound, and it diffracts around behind walls with doorways.   But λ/a is a tiny fraction for visible light passing ordinary-size objects or apertures, so light diffracts only by very small angles of direction change.

Another way of phrasing the answer: we can see by a small angle around a small obstacle or around the edge of a small opening. The side fringes in Figure 28.2 and the Arago spot in the center of Figure 28.3 show this diffraction.   We cannot always hear around corners.   Out-of-doors, away from reflecting surfaces, have someone a few meters distant face away from you and whisper. The high-frequency, short-wavelength, information-carrying components of the sound do not diffract around his head enough for you to understand her words.

9   **Question:** The diffraction grating effect is easily observed with
     everyday equipment. For example, a compact disk can be held so
     that light is reflected from it at a glancing angle (Figure
     28.28), and various colors in the reflected light can be seen.
     Furthermore, the observation depends on the orientation of the
     disk relative to the eye and light source. Explain how this
     works.

     Answer: The compact disk has many closely- and equally-spaced
     grooves. The light scattered by the grooves interferes
     constructively only in certain special directions that depend on
     the wavelength and on the direction of the incident light. Thus
     any one section of the disk functions as a diffraction grating to
     analyze white light, sending different colors off in different
     directions. Looking at the disk, you see different spectral
     colors on different sections, and the color of one section
     changes as light source, disk, or you move to change the angles
     of incidence or diffraction.

15   **Question:** How would one determine the index of refraction of a
     flat piece of dark obsidian?

     Answer: One could try to chip off a piece of the volcanic glass
     thin enough for a measurable amount of light to get through
     before being absorbed. Much more easily, use one polaroid filter
     as an analyzer to look for complete polarization of reflected
     light at Brewster's angle. The index of refraction is the
     tangent of this polarizing angle.

CHAPTER 28

PROBLEMS

28.3 **Problem:** A screen is placed 50 cm from a single slit, which is illuminated with light of wavelength 690 nm. If the distance between the first and third minima in the diffraction pattern is 3.0 mm, what is the width of the slit?

Solution: In the equation for single-slit diffraction minima at small angles, $\dfrac{y}{L} = \sin\theta = \dfrac{m\lambda}{a}$, take differences between first and third to see $\Delta y/L = \Delta m\lambda/a$ with $\Delta y = 3 \times 10^{-3}$ m, $\Delta m = 3 - 1 = 2$.

Then $a = \dfrac{\Delta m\lambda L}{\Delta y}$

$$a = \frac{(2)(690 \times 10^{-9} \text{ m})(0.5 \text{ m})}{3 \times 10^{-3} \text{ m}} = 2.3 \times 10^{-4} \text{ m } \square$$

28.5   **Problem:**   In Equation 28.5, let $\beta/2 \equiv \phi$ and show that $I = 0.5I_0$ when

$\sin \phi = \phi / \sqrt{2}.$

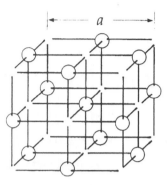

Figure 28.5

Solution: This problem along with problem 6 or problem 61 gives the half-width at half-maximum of the central peak in a single-slit diffraction pattern, which is one measure of how much a wave spreads when it goes through an opening.   Equation 28.4 becomes

$$I = I_0\left(\frac{\sin \phi}{\phi}\right)^2 = 0.5I_0 = \frac{I_0}{2}$$

$$\frac{\sin \phi}{\phi} = \frac{1}{\sqrt{2}} \qquad \sin \phi = \frac{\phi}{\sqrt{2}}$$

28.11   **Problem:**   A helium-neon laser emits light with a wavelength of 632.8 nm.   The circular aperture through which the beam emerges has a diameter of 0.50 cm.   Estimate the diameter of the beam at a distance of 10 km from the laser.

Solution: Following Equation 28.7 for diffraction from a circular opening, the beam spreads into a cone of half-angle

$$\theta_m = 1.22\frac{\lambda}{D} = 1.22\frac{(632.8 \times 10^{-9} \text{ m})}{(0.005 \text{ m})} = 1.544 \times 10^{-4} \text{ rad.}$$

The radius of the beam ten kilometers away is, from the definition of radian measure,

$r_{\text{beam}} = \theta_m(10^4 \text{ m}) = 1.544$ m, and its diameter is

$d_{\text{beam}} = 3.09$ m $\square$

28.15 **Problem:** At what distance could one theoretically distinguish two automobile headlights separated by 1.4 m? Assume a pupil diameter of 6 mm and yellow headlights (λ = 580 nm). The index of refraction in the eye is approximately 1.33.

Solution: Light from each source diffracts as it passes through the pupil of the eye. When barely resolved, the central maximum of one diffraction pattern on the retina will fall on the first minimum in the other diffraction pattern, the angle between them being $\theta = \dfrac{d}{L} = 1.22\dfrac{\lambda}{D}$

where d = separation of headlights

L = distance to car

$\lambda = \left(\dfrac{580}{1.33}\right)$ nm = wavelength of light in eye

$D = 0.006$ m

Then $L = \dfrac{Dd}{1.22\lambda}$ = 15.7 km $\square$

In air the distance to barely-resolved headlights will be much shorter because of what astronomers call seeing. The variable-density atmosphere refracts the light to change its direction, to shake the images, and at long distances, to make the headlights twinkle.

28.17 **Problem:** The Impressionist painter Georges Seurat created paintings with enormous numbers of dots of pure pigment, about 2 mm in diameter. The idea was to put colors such as red and green next to each other to form a scintillating canvas. Outside what distance would one be *un*able to discern individual dots on the canvas? (Assume λ = 500 nm within the eye and a pupil diameter of 4 mm.)

Solution: By Rayleigh's criterion, two dots separated center-to-center by 2 mm would be seen to overlap when

$$\theta = \frac{d}{L} = 1.22 \frac{\lambda}{D}$$

where d = 2 mm, L = ?, λ = 500 nm, and D = 4 mm

$$L = \frac{Dd}{1.22\lambda} = \frac{8 \times 10^{-6}}{6.1 \times 10^{-7}} = 13.1 \text{ m } \square$$

28.25 **Problem:** A source emits light with wavelengths of 531.62 nm and 531.81 nm. (a) What is the minimum number of lines required for a grating that resolves the two wavelengths in the first-order spectrum? (b) Determine the slit spacing for a grating 1.32 cm wide that has the required minimum number of lines.

Solution: The resolving power of the diffraction grating is

(a) $Nm = \dfrac{\lambda}{\Delta\lambda}$ $N(1) = \dfrac{531.7 \text{ nm}}{0.19 \text{ nm}} = 2800 \text{ lines } \square$

(b) $\dfrac{1.32 \times 10^{-2} \text{ m}}{2800} = 4.72 \text{ mm } \square$

28.27 **Problem:** White light is spread out into spectral hues by a diffraction grating. If the grating has 2000 lines per cm, at what angle will red light ($\lambda = 640$ nm) appear in first order?

Solution: The grating spacing is
$d = (10^{-2}$ m$)/2000 = 5 \times 10^{-6}$ m

Then $\sin \theta = \dfrac{m\lambda}{d} = \dfrac{1(640 \times 10^{-9}$ m$)}{5 \times 10^{-6}$ m$} = 0.128$ $\theta = 7.35°$ □

28.35 **Problem:** If an interplanar spacing of NaCl is 0.281 nm, what is the predicted angle at which x-rays of wavelength 0.14 nm will be diffracted in a first-order maximum?

Solution: The atomic planes in this crystal are shown in Figure 28.19. The diffraction they produce is described by the Bragg condition $2d \sin \theta = m\lambda$,

$\sin \theta = m\lambda/2d = (1 \times 0.14$ nm$)/(2 \times 0.281$ nm$) = 0.249$

$\theta = 14.4°$ □

28.41 **Problem:** Unpolarized light passes through two polaroid sheets. The axis of the first is vertical, and that of the second is at 30° to the vertical. What fraction of the initial light is transmitted?

Solution: The first polarizer transmits ½ the light. The second transmits $\cos^2 30° = ¾$.

$I = \dfrac{1}{2} \times \dfrac{3}{4} I_0 = \dfrac{3}{8} I_0$

28.47 **Problem:** If the polarizing angle for cubic zirconia (ZrO_2) is 65.6°, what is the index of refraction of this material?

Solution: $n = \tan \theta_p = \tan(65.6°) = 2.20$ □

28.49 **Problem:** The hydrogen spectrum has a red line at 656 nm and a blue line at 434 nm. What is the angular separation between two spectral lines obtained if the light is normally incident on a diffraction grating with 4500 lines/cm?

Solution: The grating spacing is $d = (10^{-2}$ m)/4500 = 2.22 x 10^{-6} m. In the first-order spectrum the angles of diffraction are

$$\sin \theta_1 = \frac{\lambda_1}{d}, \ \sin q_2 = \frac{\lambda_2}{d}$$

$\sin \theta_1 = (656 \times 10^{-9})/(2.22 \times 10^{-6}) = 0.295 \qquad \theta_1 = 17.17$

$\sin \theta_2 = (434 \times 10^{-9})/(2.22 \times 10^{-6}) = 0.195 \qquad \theta_2 = 11.26°$

So the separation is

$\theta_2 - \theta_1 = 5.91°$ □

In the second-order spectrum, we proceed similarly,

$\Delta\theta = \text{Arcsin}(2\lambda_1/d) - \text{Arcsin}(2\lambda_2/d) = 13.2°$ □

And in third order $\Delta\theta = 26.5°$ □

The red line does not appear in the fourth-order spectrum, so the answer is complete.

28.53 **Problem:** A diffraction grating of length 4 cm contains 6000 rulings over a width of 2 cm. (a) What is the resolving power of this grating in the first three orders? (b) If two monochromatic waves incident on this grating have a mean wavelength of 400 nm, what is their wavelength separation if they are just resolved in the third order?

Solution:

(a) From Eq. 28.10,

$R = mN$ where $N = $ (6000 lines/2 cm)(4 cm) $= 12 \times 10^3$ lines.

In first order; $R = (1)(12000) = 12000$ ☐

In second order; $R = (2)(12000) = 24000$ ☐

In third order; $R = (3)(12000) = 36000$ ☐

(b) From Eq. 28.9, $R = \dfrac{\lambda}{\Delta\lambda}$ and in third order

$$\Delta\lambda = \frac{\lambda}{R} = \frac{4.0 \times 10^{-7} \text{ m}}{3.6 \times 10^4} = 1.11 \times 10^{-11} \text{ m} = 0.0111 \text{ nm} \ \square$$

28.61 **Problem:** Another method of solving the equation $\phi = \sqrt{2} \sin \phi$ in Problem 6 is to use a scientific calculator; guess a first value of ϕ, see if it fits, and continue to update your estimate until the equation balances. How many steps (iterations) did this take? [Another approach is to apply the Newton-Raphson method to find the roots of $f(\phi) = \phi - \sqrt{2} \sin \phi$. In this approach, $\phi_2 = \phi_1 - f(\phi_1)/f'(\phi_1)$, where f' is the first derivative of f.]

Solution: We can list each trial as we try to home in on the solution to $\phi = \sqrt{2} \sin \phi$ by narrowing the range in which it must lie:

ϕ	$\sqrt{2} \sin \phi$	
1	1.19	bigger than ϕ
2	1.29	smaller than ϕ
1.5	1.41	smaller
1.4	1.394	
1.39	1.391	bigger
1.395	1.392	
1.392	1.3917	smaller
1.3915	1.39154	bigger
1.39152	1.39155	bigger
1.3916	1.391568	smaller
1.39158	1.391563	
1.39157	1.391560	
1.39156	1.391558	
1.391559	1.3915578	
1.391558	1.3915575	
1.391557	1.3915573	
1.3915574	1.3915574	

28.61 (cont)

We get the answer to seven digits after 17 steps. Clever guessing, like just using the value of $\sqrt{2}\sin\phi$ as the next guess for ϕ, could reduce this to around 13 steps. Looking back at problem 28.5, we see that the half-width at half-maximum of the central peak in a single-slit diffraction pattern is described by $1.39 = \phi = \beta/2 = \pi a \sin\theta/\lambda$

$\sin\theta = 0.443\,\lambda/a$, so the full angular width at half-maximum is $2\,\text{Arcsin}\,(0.443\lambda/a)$.

CHAPTER 29

QUESTIONS

6   **Question:** In the photoelectric effect, explain why the stopping potential depends on the frequency of light but not on the intensity.

Answer: We picture light of a higher frequency as a stream of photons of higher energy. Sometimes one photon will give essentially all of its energy to a single electron. The kinetic energy of such an electron is measured by the stopping potential. So this reverse voltage required to stop the current is proportional to the frequency of the incoming light.

More intense light consists of more photons landing on each square centimeter each second, but atoms are so small that one emitted electron never gets a kick from more than one photon. Increasing the light intensity will generally increase the size of the current, but will not change the energy of individual electrons. Thus the stopping voltage stays constant.

14   **Question:** Why is an electron microscope more suitable than an optical microscope for "seeing" objects of an atomic size?

Answer: A microscope can see details no smaller than the wavelength of the waves it images. Electrons with kinetic energies of several electronvolts have wavelengths less than a nanometer, so an electron microscope can resolve detail of this size. Visible light has wavelengths a thousand times larger.

19 **Question:** Figure 29.24 shows the spectrum of light emitted by a firefly. Determine the temperature of a black body that would emit radiation peaked at the same frequency. Based on your result, would you say firefly radiation is black-body radiation?

Answer: The wavelength emitted most intensely is 570 nm. If Wien's law applied, the temperature would be

$T = (0.2898 \times 10^{-2} \text{ m} \cdot \text{K})/(570 \times 10^{-9} \text{ m}) = 5100 \text{ K}$

Firefly radiation is not black body radiation, as the insect is made of molecules in solid and liquid phases, not of the plasma into which they would be vaporized, dissociated, and ionized at this temperature. Further, the spectral distribution of the radiation does not match a black body, which would emit much more infrared light than does the firefly.

CHAPTER 29

PROBLEMS

29.1   **Problem:** Calculate the energy of a photon whose frequency is
(a) 6.2×10^{14} Hz;   (b) 3.1 GHz;   (c) 46 MHz. Express your
answers in electron volts.

Solution: $E = hf$
(a) $E = (6.63 \times 10^{-34})(6.2 \times 10^{14}) = 4.11 \times 10^{-19}$ J

$E = 2.57$ eV □

(b) $E = (6.63 \times 10^{-34})(3.1 \times 10^{9}) = 2.06 \times 10^{-24}$ J

$E = 12.8$ meV □

(c) $E = (6.63 \times 10^{-34})(46 \times 10^{6}) = 3.05 \times 10^{-26}$ J

$E = 1.91 \times 10^{-7}$ eV □

On the electromagnetic spectrum, these are

(a)   blue light (wavelength $\lambda = c/f = 484$ nm);

(b)   a microwave with wavelength 9.67 cm; and

(c)   a radio wave in the public service LO band, in between CB
radio and television channel 2.

29.7   **Problem:** The human eye is most sensitive to light with a wavelength of λ = 560 nm. A black body of what temperature would radiate most intensely at this wavelength?

Solution: We use Wien's law

$$\lambda_{max} \, T = 0.290 \times 10^{-2} \text{ m} \cdot \text{K}$$

$$T = \frac{2.90 \text{ mm} \cdot \text{K}}{560 \times 10^{-6} \text{ mm}} = 5180 \text{ K} \; \square$$

This is close to the temperature of the surface of the sun (which is, incidentally, pretty nearly a blackbody). Living things on Earth evolved having sensitivity to electromagnetic waves around this wavelength because there is such a lot of it bouncing around carrying information.

29.11 **Problem:** Molybdenum has a work function of 4.2 eV. (a) Find the cutoff wavelength and threshold frequency for the photoelectric effect. (b) Calculate the stopping potential if the incident light has a wavelength of 180 nm.

Solution: We write Einstein's equation as $eV_s = hf - \phi$.
(a) The cutoff wavelength and threshold frequency describe light barely able to produce photoelectrons, with kinetic energy zero:

$$0 = hf_c - \phi = hc/\lambda_c - \phi$$

$$\lambda_c = \frac{hc}{\phi} = \frac{(6.63 \times 10^{-34} \text{ J} \cdot \text{s})(3 \times 10^8 \text{ m/ s})}{(4.2 \text{ eV})(1.602 \times 10^{-19} \text{ J/ eV})} = 296 \text{ nm} \; \square$$

$$f = \frac{c}{\lambda} = \frac{3 \times 10^8 \text{ m/ s}}{296 \times 10^{-9} \text{ m}} = 1.01 \times 10^{15} \text{ Hz} \; \square$$

(b) $\dfrac{hc}{\lambda} = \phi + eV_s$

$$\frac{\left(6.63 \times 10^{-34}\right)\left(3 \times 10^8\right)}{180 \times 10^{-9}}$$

$$= (4.2 \text{ eV})(1.602 \times 10^{-19} \text{ J/eV}) + (1.602 \times 10^{-19})V_s$$

$$V_s = 2.71 \text{ V} \; \square$$

29.17 **Problem:** Consider the metals lithium, beryllium, and mercury, which have work functions of 2.3 eV, 3.9 eV, and 4.5 eV, respectively. If light of wavelength 400 nm is incident on each of these metals, determine (a) which metals exhibit the photoelectric effect and (b) the maximum kinetic energy for the photoelectron in each case.

Solution: Violet light of wavelength 400 nm is a stream of photons with energy

$hf = hc/\lambda = (6.63 \times 10^{-34}$ J $\cdot$ s$)(3 \times 10^{8}$ m/s$)/(4 \times 10^{-7}$ m$)$

$hf = 4.97 \times 10^{-19}$ J$(1$ eV$/1.60 \times 10^{-19}$ J$) = 3.11$ eV

(a)   This is inadequate to supply 3.9 eV or 4.5 eV, so only lithium will show the photoelectric effect. □

(b)   For lithium $hf = \phi + K_{max}$ gives

   3.11 eV $= 2.3$ eV $+ K_{max}$ $K_{max} = 0.808$ eV □

29.19 **Problem:** Calculate the energy and momentum of a photon of wavelength 700 nm.

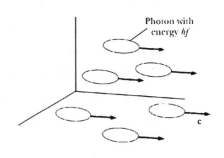

Figure 29.19

Solution: For a photon

$$E = hf = \frac{hc}{\lambda} = \frac{(6.63 \times 10^{-34})(3 \times 10^{8})}{700 \times 10^{-9}}$$

$$E = 2.84 \times 10^{-19} \text{ J}$$

$$E = 1.78 \text{ eV} \ \square$$

$$p = \frac{E}{c} = \frac{h}{\lambda} = \frac{6.63 \times 10^{-34}}{700 \times 10^{-9}}$$

$$p = 9.47 \times 10^{-28} \text{ kg} \cdot \text{m/s} \ \square$$

Notice that a photon has no mass, no charge. Its speed in vacuum is always c.

Suppose the problem asked for the energy and momentum of an electron with $\lambda = 700$ nm. The momentum would still be $p = h/\lambda = 9.47 \times 10^{-28}$ kg $\cdot$ m/s, according to deBroglie. For the electron we can use $p = mv$ to find its speed, necessarily less than c:

$$v = \frac{p}{m} = \frac{9.47 \times 10^{-28} \text{ kg} \cdot \text{m/ s}}{9.11 \times 10^{-31} \text{ kg}} = 1.04 \text{ km/s}$$

and the electron's kinetic energy is

$$\tfrac{1}{2}mv^2 = \tfrac{1}{2}(9.11 \times 10^{-31} \text{ kg})(1.04 \times 10^{3} \text{ m/s})^2$$

$$= 4.92 \times 10^{-25} \text{ J} = 3.08 \text{ meV}$$

or $\tfrac{1}{2}mv^2 = \dfrac{m^2v^2}{2m} = \dfrac{p^2}{2m}$

gives the answer without finding the speed first.

Suppose the problem asked for the energy and momentum of a proton with $\lambda = 700$ nm. Figure out yourself that $E = 2.69 \times 10^{-28}$ J $= 1.68$ neV and $p = 9.47 \times 10^{-28}$ kg $\cdot$ m/s.

29.23 **Problem:** A metal target is placed in a beam of 662 keV gamma rays emitted by a radioactive isotope of cesium (^{137}Cs). Find the energy of those photons that are scattered through an angle of 90°. The electrons in the target may be considered as free electrons.

Solution: The wavelength of the incoming gamma-rays is, from $E = hf = hc/\lambda$

$$\lambda_i = \frac{hc}{E} = \frac{\left(6.626 \times 10^{-34}\right)\left(3 \times 10^{8}\right)}{\left(662 \times 10^{3}\right)\left(1.6 \times 10^{-19}\right)} \text{ m} = 1.873 \times 10^{-12} \text{ m}$$

The Compton wavelength shift is

$$\Delta\lambda = \frac{h}{mc}(1 - \cos\theta) = \frac{6.626 \times 10^{-34} \text{ kg} \cdot \text{m}^2 / \text{s}}{(9.11 \times 10^{-31} \text{ kg})(3 \times 10^{8} \text{ m/ s})}(1 - 0)$$

$\Delta\lambda = 2.426 \times 10^{-12}$ m, to lower the photon energy by lengthening the gamma wavelength to $\lambda_f = 4.299 \times 10^{-12}$ m, and

$$E_\gamma = \frac{hc}{\lambda_f} = 288 \text{ keV} \ \square$$

The missing 662 - 288 = 374 keV is the energy given to the electrons with which the gamma rays collided.

29.27 **Problem:** Calculate the de Broglie wavelength for an electron with kinetic energy of (a) 50 eV; (b) 50 keV.

Solution: (a) We can find the electron speed first, and then the wavelength:

$$K = \tfrac{1}{2}mv^2 = (50 \text{ eV})(1.6 \text{ x } 10^{-19} \text{ C}/1 \text{ e})(1 \text{ J/C} \cdot \text{V})$$

$$v = \sqrt{2(80 \text{ x } 10^{-19} \text{ J}) / (9.11 \text{ x } 10^{-31} \text{ kg})} = 4.19 \text{ x } 10^6 \text{ m/s}$$

$$\lambda = h/p = h/mv$$

$$\lambda = (6.63 \text{ x } 10^{-34} \text{ J} \cdot \text{s}) / (9.11 \text{ x } 10^{-31} \text{ kg})(4.19 \text{ x } 10^6 \text{ m/s})$$

$$\lambda = 0.174 \text{ nm } \square$$

(b) Or we can skip computing the speed by using

$$K = mv^2/2 = p^2/2m = 50{,}000(1.6 \text{ x } 10^{-19} \text{ J})$$

$$p = 1.20 \text{ x } 10^{-22} \text{ kg} \cdot \text{m/s}$$

$$\lambda = \frac{h}{p} = 5.5 \text{ x } 10^{-12} \text{ m } \square$$

<u>Make sure</u> you can also find the wavelength of a photon when given its energy. Check that a 50 eV photon of ultraviolet light has $\lambda = hc/E = 24.9$ nm.

29.33 **Problem:** In order for an electron to be confined to a nucleus, its de Broglie wavelength has to be less than 10^{-14} m.
(a) What would be the kinetic energy of an electron confined to this region? (b) On the basis of this result, would you expect to find an electron in a nucleus? Explain.

Solution:

(a) In this problem, the electron must be treated relativistically. The momentum of the electron is not equal to mv; but it is still

$$p = \frac{h}{\lambda} = \frac{6.626 \times 10^{-34} \text{ J} \cdot \text{s}}{10^{-14} \text{ m}} = 6.626 \times 10^{-20} \text{ kg} \cdot \text{m/s}$$

The energy of the electron is

$$E = (p^2 c^2 + m^2 c^4)^{1/2}$$

$$E = [(6.626 \times 10^{-20})^2 (3 \times 10^8)^2$$
$$+ (0.511 \times 10^6)^2 (1.6 \times 10^{-19})^2]^{1/2}$$

$$E = 1.99 \times 10^{-11} \text{ J} = 1.24 \times 10^8 \text{ eV}$$

so that $K = E - m_e c^2 \approx 124$ MeV □

(b) The kinetic energy is too large to expect that the electron could be confined to a region the size of the nucleus. The electric potential energy of an electron distant by 10^{-14} m from a proton is

$$U = qV = -ek_e e/r$$

$$U = -e(8.99 \times 10^9 \text{ N} \cdot \text{m}^2/\text{C}^2)(1.6 \times 10^{-19} \text{ C})/(10^{-14} \text{ m})$$

$$U = -0.144 \text{ MeV}.$$

The kinetic energy of +124 MeV would make the electron immediately escape the proton's attraction.

29.35 Problem: Robert Hofstadter won the 1961 Nobel Prize in physics for his pioneering work in scattering 20-GeV electrons from nuclei.   (a)   What is the γ-factor for a 20-GeV electron, where $\gamma = (1 - v^2/c^2)^{-1/2}$?   What is the momentum of the electron in kg · m/s?   (b) What is the wavelength of a 20-GeV electron, and how does it compare with the size of a nucleus?

Solution:

(a)   From $E = \gamma m_e c^2$ (Eq. 10.21)

$$\gamma = \frac{20 \times 10^3 \text{ MeV}}{0.511 \text{ MeV}} = 39,139 \ \square$$

For these extreme-relativistic electrons, with $m_e c^2 \ll pc$

$E^2 = p^2 c^2 + m^2 c^4$ simplifies to $E^2 \cong p^2 c^2$

$$p = \frac{E}{c} = \frac{(2 \times 10^4 \text{ MeV})(1.6 \times 10^{-13} \text{ J/ MeV})}{3 \times 10^8 \text{ m/ s}}$$

$p = 1.07 \times 10^{-17} \text{ kg} \cdot \text{m/s} \ \square$

(b) $\lambda = \dfrac{h}{p} = \dfrac{6.626 \times 10^{-34} \text{ J} \cdot \text{s}}{1.07 \times 10^{-17} \text{ kg} \cdot \text{m/ s}} \cong 6.2 \times 10^{-17} \text{ m} \ \square$

Since the size of a nucleus is on the order of 10^{-14} m, the 20-GeV electrons would be small enough to go through the nucleus.

29.39 **Problem:** A neutron beam with a selected speed of 0.4 m/s is directed through a double slit with a 1-mm separation. An array of detectors is placed 10 m from the slit. (a) What is the de Broglie wavelength of the neutrons? (b) How far off axis is the first zero-intensity point on the detector array? (c) Can we say which slit the neutron passed through? Explain.

Solution: (a) $\lambda = \dfrac{h}{mv} = \dfrac{6.63 \times 10^{-34}}{(1.67 \times 10^{-27})(0.4)} = 9.93 \times 10^{-7}$ m $\square$

(b) The condition for destructive interference in a multiple-slit experiment is $d \sin \theta = (m + \frac{1}{2})\lambda$ with $m = 0$ for the first minimum. Then,

$$\theta = \sin^{-1}\left(\frac{\lambda}{2d}\right) = 0.0284° \qquad \frac{y}{D} = \tan \theta$$

$$y = D \tan \theta = (10 \text{ m})(\tan 0.0284°) = 4.97 \text{ mm} \ \square$$

(c) We cannot say the neutron passed through one slit. We can only say it passed through the pair of slits, as a water wave does to produce an interference pattern.

29.43 **Problem:** An electron (m = 9.11 x 10^{-31} kg) and a bullet (m = 0.02 kg) each have a speed of 500 m/s, accurate to within 0.01%. Within what limits could we determine the position of the paths of the objects?

Solution: For the electron the uncertainty in momentum is

$\Delta p = m\Delta v = (9.11 \times 10^{-31}$ kg$)(500$ m/s$)(10^{-4})$

$\Delta p = 4.56 \times 10^{-32}$ kg $\cdot$ m/s

The minimum uncertainty in position is then

$$\Delta x = \frac{h}{4\pi\Delta p} = \frac{6.63 \times 10^{-34} \text{ J}\cdot\text{s}}{4\pi (4.56 \times 10^{-32} \text{ kg}\cdot\text{m/ s})} = 1.16 \text{ mm} \ \square$$

For the bullet

$\Delta p = m\Delta v = (0.02$ kg$)(500$ m/s$)(10^{-4}) = 10^{-3}$ kg $\cdot$ m/s

$$\Delta x = \frac{h}{4\pi\Delta p} = 5.28 \times 10^{-32} \text{ m} \ \square$$

Quantum mechanics describes all objects, but the quantum fuzziness in position is unobservably small for the bullet and large for the small-mass electron.

CHAPTER 30

QUESTIONS

2   **Question:** Must an atom first be ionized before it can emit
     light? Discuss.

     Answer: An atom must absorb energy to shift to an excited state
     before it can emit light, but it need not be ionized.

5   **Question:** When a hologram is produced, the system (including
     light source, object, beam splitter, and so on) must be held
     motionless within a quarter of a wavelength. Why?

     Answer: The hologram is an interference pattern between light
     scattered from the object and the reference beam. If anything
     moves by a distance comparable to the wavelength of the light, or
     more, the pattern will wash out. The effect is just like making
     the slits vibrate in Young's experiment.

14   **Question:** It is easy to understand how two electrons (one spin
     up, one spin down) can fill the $1s$ shell for a helium atom. How
     is it possible that eight more electrons can fit into the $2s$, $2p$
     levels to complete the $1s^2 2s^2 2p^6$ shell for a neon atom?

     Answer: Each of the eight electrons must have just one quantum
     number different from each of the others. They can differ (in
     m_s) by being spin-up or spin-down. They can also differ (in ℓ)
     in angular momentum and in the general shape of the wave function
     (look at the $2s$ and $2p$ graphs in Figure 30.8). Those electrons
     with $\ell = 1$ can differ (in m_ℓ) in orientation of angular momentum
     -- look at Figure 30.9.

17   **Question:**   The efficiencies of most solid-state lasers are on the order of 1 to 2%. Although the laser output is monochromatic and highly directional, can you use Figures 30.22 and 30.23 to determine why the energy input must exceed laser energy output by a factor of 50 to 100?

Answer: Only energy in a narrow spectral band from the flashlamp will cause transitions from E_1 to E_3. Additionally, not all the light emitted by the flashlamp is absorbed in the ruby rod.

CHAPTER 30

PROBLEMS

30.3 **Problem:** A general expression for the energy levels of one-electron atoms is $E_n = -\left(\dfrac{\mu k_e^2 q_1^2 q_2^2}{2\hbar^2}\right)\dfrac{1}{n^2}$ where k_e is the Coulomb constant, q_1 and q_2 are the charges of the two particles, and m is the reduced mass given by $\mu = m_1 m_2/(m_1 + m_2)$. In Problem 2 we found that the wavelength for the $n = 3$ to $n = 2$ transition of the hydrogen atom is 656.3 nm (visible red light). What are the wavelengths for this same transition in (a) positronium, which consists of an electron and a positron, and (b) singly ionized helium? (*Note:* A positron is a positively charged electron.)

Solution: For hydrogen, $\mu = (m_{proton} m_e)/(m_{proton} + m_e) \cong m_e$.
The photon energy is $E_3 - E_2$ and its wavelength is $\lambda = c/f = hc/(E_3 - E_2) = 656.3$ nm

(a) For positronium, $\mu = (m_e m_e)/(m_e + m_e) = m_e/2$, so the energy of each level is one-half as large as in hydrogen (protonium). The photon energy is inversely proportional to its energy, so $\lambda_{32} = 2(656 \text{ nm}) = 1312$ nm ☐ which is in the infrared region.

(b) For He$^+$, $\mu \approx m_e$, $q_1 = e$, and $q_2 = 2e$, so each energy is $2^2 = 4$ times larger than hydrogen.

Then $\lambda_{32} = \left(\dfrac{656}{4}\right)$ nm = 164 nm ☐

which is the ultra-violet region.

30.11 **Problem:** If a muon (a negatively charged particle with mass
       206 times the electron's mass) is captured by a lead nucleus,
       $Z = 82$, the resulting system behaves like a one-electron atom.
       (a) What is the "Bohr radius" for a muon captured by a lead
       nucleus? (*Hint:* Use Eq. 30.4.) (b) Using Equation 30.2 with e
       replaced by Ze, calculate the ground-state energy of a muon
       captured by a lead nucleus. (c) What is the transition energy
       of a muon descending from the $n = 2$ to the $n = 1$ level in a
       muonic lead atom?

   Solution:

   (a) We expand the treatment of the Bohr atom in Section 12.5
        to particle of charge $-e$ and mass m orbiting a much more
        massive nucleus of charge Ze. Angular momentum is
        quantized according to $mvr = nh$

        $n = 1, 2, 3, \ldots$ and $\Sigma F = ma$ means $k_e Ze^2/r^2 = mv^2/r$.

        We eliminate v with $v = n\hbar/mr$;

        $k_e Ze^2/r^2 = mn^2\hbar^2/m^2 r^3$

        $r = n^2 \hbar^2/mk_e Ze^2$

        So the generalized Bohr radius is $\hbar^2/mkZe^2$.

        With $m = 206\ m_e$ and $Z = 82$, the radius for muonic lead is

        $a_\mu = a_0/(206 \times 82) = 3.13$ fm $\square$

   (b) The energy is

        $\tfrac{1}{2}mv^2 - k_e Ze^2/r = \tfrac{1}{2}k_e Ze^2/r - k_e Ze^2/r = -k_e Ze^2/2r$

        $\qquad = -k_e Ze^2/2(n^2\hbar^2/mk_e Ze^2) = -mk_e^2 Z^2 e^4/2\hbar^2 n^2$

        For the ground-state energy we take $n = 1$:

        $E_{\mu 1} = -mk_e^2 Z^2 e^4/2\hbar^2 = -206 \times 82^2 \times 13.6$ eV $= -18.8$ MeV $\square$

   (c) $E_{\mu 2} = \dfrac{-18.8}{4} = -4.7$ MeV

        $\Delta E_{\mu 2 \,\to\, \mu 1} = 14.1$ MeV $\square$

30.13 **Problem:** If an electron has an orbital angular momentum of 4.714×10^{-34} J · s, what is the orbital quantum number for this state of the electron?

Solution: $L = \sqrt{\ell(\ell + 1)}\hbar$

$$4.714 \times 10^{-34} = \sqrt{\ell(\ell + 1)} \frac{6.63 \times 10^{-34}}{2\pi}$$

$$\ell(\ell + 1) = \frac{4.714 \times 10^{-34})^2 (2\pi)^2}{(6.63 \times 10^{-34})^2} = 1.996 \times 10^1 \approx 20 = 4(4 + 1)$$

So $\ell = 4$ □

30.15 **Problem:** How many different sets of quantum numbers are possible for an electron for which (a) $n = 1$? (b) $n = 2$? (c) $n = 3$? (d) $n = 4$? (e) $n = 5$? Check your results to show that they agree with the general rule that the number of different sets of quantum numbers is equal to $2n^2$.

Solution:

(a) $n = 1$: For $n = 1$, $\ell = 0$, $m\ell = 0$, $m_s = \pm\frac{1}{2}$, ® 2 sets

n	ℓ	$m\ell$	m_s
1	0	0	$-\frac{1}{2}$
1	0	0	$+\frac{1}{2}$

$2n^2 = 1(1)^2 = 2$ □

(b) For $n = 2$, we have

n	ℓ	$m\ell$	m_s	
2	0	0	$\pm\frac{1}{2}$	
2	1	-1	$\pm\frac{1}{2}$	yields 8 sets;
2	1	0	$\pm\frac{1}{2}$	$2n^2 = 2(2)^2 = 8$ □
2	1	1	$\pm\frac{1}{2}$	

Note that the number is twice the number of m_1 values. Also, for each ℓ there are $(2\ell + 1)$ m_1 values. Finally, ℓ can take on values ranging from 0 to $n - 1$. So the general expression is $s = \sum_{0}^{n-1} 2(2\ell + 1)$. The series is an arithmetic progression $2 + 6 + 10 + 14$, the sum of which is

$$s = \frac{n}{2}\left[2a + (n - 1)d\right] \text{ where } a = 2,\ d = 4.$$

$$s = \frac{n}{2}\left[4 + (n - 1)4\right]$$

(c) $n = 3$: $2(1) + 2(3) + 2(5) = 2 + 6 + 10 = 18$

$2n^2 = 2(3)^2 = 18$ □

30.15 (cont)

(d) $n = 4$: $2(1) + 2(3) + 2(5) + 2(7) = 32$

$2n^2 = 2(4)^2 = 32$ ☐

(e) $n = 5$: $32 + 2(9) = 32 + 18 = 50$ $2n^2 = 2(5)^2 = 50$ ☐

30.25   **Problem:**   (a) Scanning through Table 30.4 in order of increasing atomic number, note that the electrons fill the subshells in such a way that those subshells with the lowest values of $n + \ell$ are filled first. If two subshells have the same value of $n + \ell$, the one with the lower value of n is filled first. Using these two rules, write the order in which the subshells are filled through $n = 7$. (b) Predict the chemical valence for elements with atomic numbers 15, 47, and 86, and compare them with the actual valences.

Solution: (a)

$n + \ell$	1	2	3	4	5	6	7
subshell	1s	2s	2p,3s	3p,4s	3d,4p,5s	4d,5p,6s	4f,5d,6p,7s

(b)   Z = 15: Filled subshells: 1s,2s,2p,3s (12 electrons)

Valence subshell: 3 electrons in 3p subshell

Prediction: Valence = +3 or -5

Element is Phosphorus = valence +3 or -5

Z = 47: Filled subshells: 1s, 2s, 2p, 3s, 3p, 4s, 3d, 4p, 5s (38 electrons)

Outer subshell: 9 electrons in 4d subshell

Prediction: Valence = -1

Element is Silver,   (Prediction fails)

valence +1.

Z = 86: Filled shells: 1s, 2s, 2p, 3s, 3p, 4s, 3d, 4p, 5s, 4d, 5p, 6s, 4f, 5d, 6p

Outer subshell is full--predict inert gas

Element is Radon, inert.

30.29 **Problem:** Use the method illustrated in Example 30.7 to calculate the wavelength of the x-ray emitted from a molybdenum target ($Z = 42$) when an electron undergoes a transition from the L shell ($n = 2$) to the K shell ($n = 1$).

Solution: Following Example 30.7, we suppose the electron is originally in the L shell with just one other electron in the K shell between it and the nucleus, so it moves in a field of effective charge $(42 - 1)e$. Its energy is then $E_L = -(42 - 1)^2 13.6$ eV/4. In its final state we estimate the screened charge holding it in orbit as again $(42 - 1)e$, so its energy is $E_K = -(42 - 1)^2 13.6$ eV. The photon energy emitted is the difference.

$$E_\gamma = \frac{3}{4}(42 - 1)^2(13.6 \text{ eV}) = 1.71 \times 10^4 \text{ eV} = 2.74 \times 10^{-15} \text{ J.}$$

Then $f = \dfrac{E}{h} = 4.14 \times 10^{18}$ Hz

$$\lambda = \frac{c}{f} = 0.725 \text{ Å} \quad \square$$

30.33 **Problem:** A ruby laser delivers a 10-ns pulse of 1 MW average power. If all the photons are of wavelength 694.3 nm, how many photons are contained in the pulse?

Solution: The energy of the pulse is
$E = (10^6 \text{ W})(10^{-8} \text{ s}) = 10^{-2}$ J.
The energy of each photon in the pulse is

$$E_\gamma = hf = \frac{hc}{\lambda} = \frac{(6.626 \times 10^{-34})(3 \times 10^8)}{694.3 \times 10^{-9}} \text{ J} = 2.86 \times 10^{-19} \text{ J.}$$

So $N = \dfrac{E}{E_\gamma} = \dfrac{10^{-2}}{2.86 \times 10^{-19}} = 3.5 \times 10^{16}$ photons $\quad \square$

30.37 **Problem:** In the technique known as electron spin resonance
(ESR), a sample containing unpaired electrons is placed in a
magnetic field. Consider the simplest situation, that in which
there is only one electron and therefore only two possible
energy states exist, corresponding to $m_s = \pm\frac{1}{2}$. In ESR, the
electron's spin magnetic moment is "flipped" from a lower
energy state to a higher energy state by the absorption of a
photon. (The lower energy state corresponds to the case where
the magnetic moment μ_s is aligned with the magnetic field, and
the higher energy state corresponds to the case where μ_s is
aligned against the field.) What is the photon frequency
required to excite an ESR transition in a magnetic field of
0.35 T?

Solution: As in section 19.4, the magnetic moment feels torque
$\tau = \mu \times \mathbf{B}$ in an external field. In turning it from alignment
with the field to the opposite direction, the field does work
according to Equation 11.32,

$$W = \int dW = \int_0^{180°} \tau d\theta = \int_0^{\pi} \mu B \sin\theta d\theta$$

$$W = \mu B \cos\theta \big|_0^{\pi} = 2 \ mB.$$

The photon must carry this much energy to make the electron
flip:

$$\Delta E = 2\mu_B B = hf$$

$$2(9.27 \times 10^{-24})(0.35) = (6.63 \times 10^{-34})f$$

so $f = 9.79 \times 10^9$ Hz □

30.45 **Problem:** *Positronium* is a hydrogen-like atom consisting of a positron (a positively charged electron) and an electron revolving around each other. Using the Bohr model, find the allowed radii (relative to the center of mass of the two particles) and the allowed energies of the system.

Solution: Let r represent distance between electron and positron. The two move with equal-size and opposite velocities in the same circle, radius $r/2$, around their center of mass. Total angular momentum is quantized according to

$$mvr/2 + mvr/2 = nh, \quad n = 1, 2, 3, \ldots$$

For each $\Sigma F = ma$ reads $\dfrac{k_e e^2}{r^2} = \dfrac{mv^2}{r/2}$

We eliminate $v = n\hbar/mr$ to find

$$\frac{k_e e^2}{r} = \frac{2 m n^2 \hbar^2}{m^2 r^2}$$

$$r = \frac{2 n^2 \hbar^2}{m k_e e^2} = 2 a_0 n^2 = (1.06 \times 10^{-10} \text{ m}) n^2$$

for the allowed separation distances. $\square$

The energy is

$$\frac{1}{2} mv^2 + \frac{1}{2} mv^2 - \frac{k_e e^2}{r} = mv^2 - \frac{k_e e^2}{r}.$$

with $mv^2 = k_e e^2/2r$ from above, we have

$$E = \frac{k_e e^2}{2r} - \frac{k_e e^2}{r} = -\frac{k_e e^2}{2r} = -\frac{k_e e^2}{4 a_0 n^2}$$

This is one-half the energy of hydrogen, or

$$E = \frac{-13.6 \text{ ev}}{2 n^2} = \frac{-6.80 \text{ eV}}{n^2} \quad \square$$

30.53 **Problem:** (a) Calculate the most probable radius for an
electron in the 2s state of hydrogen. (*Hint:* Let $x = r/a_0$,
find an equation of x, and show that $x = 5.236$ is a solution to
this equation.) (b) Show that the wave function given by
Equation 30.8 is normalized.

Solution: We use $\psi_{2s}(r) = \dfrac{1}{4}(2\pi\, a_0^3)^{-1/2}\left(2 - \dfrac{r}{a_0}\right)e^{-r/2a_0}$ and by Eq.

30.6. The radial probability distribution function is

$P(r) = 4\pi r^2 \psi^2 = \dfrac{1}{8}\left(\dfrac{r^2}{a_0^3}\right)\left(2 - \dfrac{r}{a_0}\right)^2 e^{-r/a_0}$. Its extrema are given by

(a) $\dfrac{dP(r)}{dr} = \dfrac{1}{8}\left[\dfrac{2r}{a_0^3}\left(2 - \dfrac{r^2}{a_0}\right)^2 - \dfrac{2\,r^2}{a_0^3}\left(\dfrac{1}{a_0}\right)\left(2 - \dfrac{r}{a_0}\right) - \dfrac{r^2}{a_0^3}\left(2 - \dfrac{r}{a_0}\right)^2\left(\dfrac{1}{a_0}\right)\right]e^{-r/a_0} = 0$

or $\dfrac{1}{8}\left(\dfrac{r}{a_0^3}\right)\left(2 - \dfrac{r}{a_0}\right)\left[2\left(2 - \dfrac{r}{a_0}\right) - \dfrac{2r}{a_0} - \dfrac{r}{a_0}\left(2 - \dfrac{r}{a_0}\right)\right]e^{-r/a_0} = 0$

Therefore $[\ldots\ldots] = 4 - \dfrac{6r}{a_0} + \left(\dfrac{r}{a_0}\right)^2 = 0$ which has solutions

$r = (3 \pm \sqrt{5})\, a_0$

$\left[\text{The roots of } \dfrac{dP}{dr} = 0 \text{ at } r = 0, \ r = 2a_0 \text{ and } r = \infty\right.$

$\left.\text{are minima } (\psi = 0).\right]$

We substitute the two roots into $P(r)$:

When $r = (3 - \sqrt{5})\, a_0 = 0.764a_0$, then $P(r) = \dfrac{0.0519}{a_0}$

When $r = (3 + \sqrt{5})\, a_0 = 5.24a_0$, then $P(r) = \dfrac{0.191}{a_0}$

Therefore the most probable value of r is

$(3 + \sqrt{5})\, a_0 = 5.24a_0$ □

30.53 (cont.)

(b) $\int_0^\infty P(r)\, dr = \int_0^\infty \frac{1}{8}\left(\frac{r^2}{a_0^3}\right)\left(2 - \frac{r}{a_0}\right)^2 e^{-r/a_0}\, dr$

Let $u = \dfrac{r}{a_0}$, $dr = a_0 du$, then

$\int_0^\infty P(r)dr = \int_0^\infty \frac{1}{8}u^2\left(4 - 4u + u^2\right)e^{-u}\, du$

$\int_0^\infty P(r)dr = \int_0^\infty \frac{1}{8}\left(u^4 - 4u^3 + 4u^2\right)e^{-u}\, du$

Use a table of integrals or integrate by parts repeatedly to find

$\int_0^\infty P(r)dr = \left. -\frac{1}{8}\left(u^4 + 4u^2 + 8u + 8\right)e^{-u}\right|_0^\infty = 1$ as desired

30.55 **Problem:** For hydrogen in the 1s state, what is the probability of finding the electron farther than $2.50a_0$ from the nucleus?

Solution: The radial probability distribution function is $P(r) = 4\pi r^2 |\psi|^2$. With $\psi_{1s} = (\pi a_0^3)^{-1/2} e^{-r/a_0}$ it is $P(r) = 4 r^2 a_0^{-3} e^{-2r/a_0}$. The required probability is then

$P = \int_{2.5a_0}^\infty P(r)dr = \int_{2.5a_0}^\infty \frac{4 r^2}{a_0^3} e^{-2r/a_0}\, dr$

Let $z = \dfrac{2r}{a_0}$, $dz = \dfrac{2dr}{a_0}$

$P = \frac{1}{2}\int_5^\infty z^2 e^{-z}\, dz$

$P = \left. -\frac{1}{2}\left(z^2 + 2z + 2\right)e^{-z}\right|_5^\infty$

$P = -\frac{1}{2}[0] + \frac{1}{2}(25 + 10 + 2)e^{-5} = \left(\frac{37}{2}\right)(0.00674)$

$P = 0.125$ □

30.57 **Problem:** Light from a certain He-Ne laser has a power output of 1.0 mW and a cross-sectional area of 10 mm^2. The entire beam is incident on a metal target that requires 1.5 eV to remove an electron from its surface. (a) Perform a classical calculation to determine how long it would take one atom in the metal to absorb 1.5 eV from the incident beam. (*Hint:* Assume that the area of an atom is 1 Å^2 = 10^{-20} m^2, and first calculate the energy incident on each atom per second.) (b) Compare the (wrong) answer obtained in (a) to the actual response time for photo-electric emission ($\approx 10^{-9}$ s), and discuss the reasons for the large discrepancy.

Solution: The intensity of the beam is
$I = P/A = (10^{-3}$ W$)/(10 \times 10^{-6}$ m$^2) = 100$ W/m^2.
The power incident on our one atom is
$P_a = IA = (100$ W/m$^2)10^{-20}$ m$^2 = 10^{-18}$ W
To absorb 1.5 eV = 2.4×10^{-19} J, the atom must be in the beam for time $t = E/P = (2.4 \times 10^{-19}$ J$)/(10^{-18}$ J/s$) = 0.24$ s

(b) The classical answer is too large by a factor of about a billion because the beam does not carry smeared-out energy but energy in photon lumps. The atoms do not all have to wait together for energy to accumulate, like school teachers for retirement. In the first nanosecond of absorption, some atom will get hit with a photon and release an electron. The photocurrent begins right away.

CHAPTER 31

QUESTIONS

2   **Question:** Estimate the mass of a pinhead composed entirely of densely packed nuclear matter.

Answer: Suppose the pinhead is a sphere ½ mm in diameter. Its volume is then $(4/3)\pi(1/4 \times 10^{-3}$ m$)^3 \sim 7 \times 10^{-11}$ m^3. Example 31.1 shows that nuclear matter has a density 2.3×10^{17} kg/m^3, so the mass is $\rho V = (7 \times 10^{-11}$ m$)(2.3 \times 10^{17}$ kg/m$^3) \approx 10^7$ kg or ten tonnes.

8   **Question:** What fraction of a radioactive sample has decayed after two half-lives have elapsed?

Answer: Three-quarters of a radioactive sample has decayed after two half-lives. The quantity remaining is
$N = N_0 e^{-lt} = N_0 \exp(-l2T_{\frac{1}{2}}) = N_0 \exp(-l2 \times 0.693/l)$
$N = N_0 \exp(-1.39) = 0.25N_0$.

13   **Question:** If a nucleus such as ^{226}Ra that is initially at rest undergoes alpha decay, which has more kinetic energy after the decay, the alpha particle or the daughter nucleus?

Answer: The alpha particle and the daughter nucleus carry equal and opposite amounts of momentum. Since kinetic energy can be written as $p^2/2m$, the small-mass alpha particle has much more of the decay energy than the recoiling nucleus.

19 **Question:** The radioactive nucleus $^{222}_{88}$Ra has a half-life of about 1.6 x 10^3 years. Given that the Solar System is about 5 billion years old, how can you explain why we still can find this nucleus in nature?

Answer: A long-lived progenitor at the top of one of the three natural radioactive series is the source of our radium. As an example, Thorium-232 with a half-life of 14 billion years a-decays to Radium-228 (See Figure 31.16).

CHAPTER 31

PROBLEMS

31.7 **Problem:** The nucleus of an iron atom has a radius equal to 4.60×10^{-15} m. What must be the minimum speed of an α particle if it is to reach the nucleus? Disregard the effect of the outer electrons.

Solution: Assume a head-on collision for the minimum speed situation, and use conservation of energy:

$$\frac{1}{2} mv^2 = Q_2 V_1 = \frac{k_e Q_1 Q_2}{r_{min}}, \text{ thus}$$

$$v = \sqrt{\frac{2\left(8.99 \times 10^9 \text{ N} \cdot \text{m}^2 / \text{C}^2\right)(26)(2)(1.6 \times 10^{-19} \text{ C})^2}{4(1.66 \times 10^{-27} \text{ kg})(4.6 \times 10^{-15} \text{ m})}}$$

$$v = 2.80 \times 10^7 \text{ m/s} \ \square$$

31.9 **Problem:** Certain stars are thought to collapse at the ends of their lives, combining their protons and electrons to form a neutron star. Such a star could be thought of as a giant atomic nucleus. If a star with a mass equal to that of the Sun ($M = 1.99 \times 10^{30}$ kg) collapsed into neutrons ($m_n = 1.67 \times 10^{-27}$ kg), what would be the radius of such a star? (Assume that $r = r_0 A^{1/3}$.)

Solution: The number of nucleons in a star of one solar mass is

$$A = \frac{2 \times 10^{30}}{1.67 \times 10^{-27}} = 1.20 \times 10^{57}$$

Therefore

$$r = r_0 A^{1/3} = (1.2 \times 10^{-15} \text{ m})(1.06 \times 10^{19}) = 12.7 \text{ km} \ \square$$

31.15 **Problem:** Calculate the minimum energy required to remove a neutron from the $^{43}_{20}$Ca nucleus.

Solution: Removal of a neutron from $^{43}_{20}$Ca would result in the residual nucleus, $^{42}_{20}$Ca. If the required separation energy is S_n, the overall process can be described by

$\text{mass}\left(^{43}_{20}\text{Ca}\right) + S_n = \text{mass}\left(^{42}_{20}\text{Ca}\right) + \text{mass}(n)$ or

$S_n = (41.95863 + 1.008665 - 42.958770)\,\text{u}$

$S_n = (0.008525\ \text{u})(931.5\ \text{MeV/u}) = 7.94\ \text{MeV}$ $\square$

31.17 **Problem:** The $^{139}_{57}$La isotope of lanthanum is stable. A radioactive isobar (see Problem 16) of this lanthanum isotope, $^{139}_{59}$Pr is located below the line of stable nuclei in Figure 31.3 and decays by β^+ emission. Another radioactive isobar of ^{139}La, $^{139}_{55}$Cs decays by β^- emission and is located above the line of stable nuclei in Figure 31.3. (a) Which of these three isobars has the highest neutron-to proton ratio? (b) Which has the greatest binding energy per nucleon? (c) Which of the two radioactive nuclei (^{139}Pr or ^{139}Cs) do you expect to be heavier?

Solution:

(a)   For $^{139}_{59}$Pr the neutron number is 139 - 59 = 80. For $^{139}_{55}$Cs the neutron number is 84, so this cesium isotope has the greatest neutron-to-proton ratio. It really lies above the zone of stability while the praseodymium lies below.

(b)   Binding energy per nucleon measures stability so it is greatest for the stable nucleus, the lanthanum isotope. Note also that it has a magic number of neutrons, 82.

(c)   Cs-139 has 55 protons and 84 neutrons. Pr-139 has 59 protons and 80 neutrons. Since neutrons are heavier than protons (by about 1.3 MeV), we would expect that the more neutron-rich nucleus Cs-139 would be heavier. □

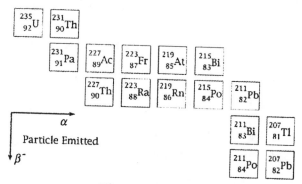

Figure 31.17

31.19 Problem: A sample of radioactive material contains 10^{15} atoms and has an activity of 6.00×10^{11} Bq. What is the half-life for this material?

Solution: $\dfrac{dN}{dt} = -\lambda N$

$\lambda = \dfrac{1}{N}\left(-\dfrac{dN}{dt}\right) = 10^{-15}\left(6.00 \times 10^{11}\ \text{s}^{-1}\right) = 6.00 \times 10^{-4}\ \text{s}^{-1}$

$T_{1/2} = \dfrac{\ln 2}{\lambda} = 1160\ \text{s}\ \square\ (= 19.3\ \text{min})$

31.27 Problem: A building has accidentally become contaminated with radioactivity. The longest-lived material in the building is strontium-90 $^{90}_{38}\text{Sr}$ atomic mass 89.9077). If the building initially contained 5.0 kg of this substance and the safe level is less than 10.0 counts/min, how long will the building be unsafe?

Solution: The number of muclei in the original sample is

$N_0 = \dfrac{\text{mass present}}{\text{mass of nucleus}} = \dfrac{5.0\ \text{kg}}{89.9077\,u(1.66 \times 10^{-27}\ \text{kg/ u})}$

$N_0 = 3.35 \times 10^{25}$ nuclei

$\lambda = \dfrac{0.693}{T_{1/2}} = \dfrac{0.693}{28.8\ \text{y}} = 2.4063 \times 10^{-2}\ \text{y}^{-1} = 4.575 \times 10^{-8}\ \text{min}^{-1}$

(half-life is taken from appendix A3)

$R_0 = \lambda N_0 = (4.575 \times 10^{-8}\ \text{min}^{-1})(3.35 \times 10^{25})$

$R_0 = 1.533 \times 10^{18}$ counts/min

$\dfrac{R}{R_0} = e^{-\lambda t} = \dfrac{10\ \text{counts/ min}}{1.533 \times 10^{18}\ \text{counts/ min}} = 6.525 \times 10^{-18}$

and, $\lambda t = -\ln(6.525 \times 10^{-18}) = 39.57$

giving $t = 1645\ \text{y}\ \square$

31.29 **Problem:** Find the energy released in the alpha decay of $^{238}_{92}\text{U}$

$$^{238}_{92}\text{U} \rightarrow {}^{234}_{90}\text{Th} + {}^{4}_{2}\text{He}$$

You will find the following mass values useful:

$$M\left({}^{238}_{92}\text{U}\right) = 238.050786 \text{ u}$$

$$M\left({}^{238}_{90}\text{Th}\right) = 234.043583 \text{ u}$$

$$M\left({}^{4}_{2}\text{He}\right) = 4.002603 \text{ u}$$

Solution: $Q = (M_u - M_{Th} - M_{He})(931.5 \text{ MeV/u})$

$Q = (238.050786 - 234.043583 - 4.002603)(931.5) = 4.28 \text{ MeV}$ □

31.35 **Problem:** Starting with $^{235}_{92}\text{U}$ the following sequence of decays is observed, ending with the stable isotope $^{207}_{82}\text{Pb}$ (Fig.31.17). Enter the correct isotope symbol in each open square.

Solution: Whenever an $\alpha = {}^{4}_{2}\text{He}$ is emitted, Z drops by 2 and A by 4. Whenever a $\beta^- = {}^{0}_{-1}\beta^-$ is emitted, Z increases by 1 and A is unchanged. We find the chemical name by looking up Z in a periodic table.

31.39 Problem: The nucleus $^{15}_{8}O$ decays by electron capture. Write (a) the basic nuclear process and (b) the decay process referring to neutral atoms.   (c) Determine the energy of the neutrino. Disregard the daughter's recoil.

Solution: (a) $e^- + p \rightarrow n + \nu$

(b)   Add 7 protons, 7 neutrons, and 7 electrons to each side to give $^{15}O \rightarrow ^{15}N + \nu$

(c)   From Table A.3,

$$m(^{15}O) = m(^{15}N) + Q/c^2$$

$$\Delta m = 15.003065 - 15.000109 = 0.002956 \text{ u}$$

$$Q = (931.5 \text{ MeV/u})(0.002956 \text{ u}) = 2.75 \text{ MeV} \;\square$$

31.41 Problem: The following is the first known reaction (achieved in 1934) in which the product nucleus is radioactive:

$$^{27}_{13}Al\,(\alpha, n)^{30}_{15}P$$

Calculate the Q value of this reaction.

Solution: $Q = [M_\alpha + M(^{27}Al) - M(^{30}P) - M_n](931.5)$

$Q = (4.002603 + 26.981541 - 29.078310 - 1.008665)(931.5)$

$Q = -2.64 \text{ MeV} \;\square$

The reaction is endoenergetic: you must put in 2.64 MeV to make it happen.

31.43 **Problem:** Natural gold has only one isotope, $^{197}_{79}$Au If natural gold is irradiated by a flux of slow neutrons, β^- particles are emitted.   (a) Write the appropriate reaction equations. (b) Calculate the maximum energy of the emitted beta particles. The mass of $^{198}_{80}$Hg is 197.96675 u.

Solution:

(a)   The Au-197 will absorb a neutron to become $^{198}_{79}$Au which emits a b$^-$ to become $^{198}_{80}$Hg. The reaction for nuclei is $^{197}_{79}$Au nucleus + $^{1}_{0}$n $\rightarrow$ $^{198}_{80}$Hg nucleus + $^{0}_{-1}\beta$ + ν. Add 79 electrons to both sides to have the reaction for neutral atoms. $^{197}_{79}$Au + $^{1}_{0}$n $\rightarrow$ $^{198}_{80}$Hg + ν

(b)   From Table A.3,

$196.96656 + 1.008665 = 197.96675 + 0 + Q/c^2$

$Q = \Delta mc^2 = (0.008475 \text{ u})(931.5 \text{ MeV/u})$

$Q = 7.89 \text{ MeV}$ □

31.55 **Problem:** A by-product of some fission reactors is the isotope $^{239}_{94}Pu$ which is an alpha emitter with a half-life of 24,000 years: $^{239}_{94}Pu \rightarrow {}^{235}_{92}U + \alpha$

Consider a sample of 1 kg of pure $^{239}_{94}Pu$ at $t = 0$. Calculate (a) the number of $^{239}_{94}Pu$ nuclei present at $t = 0$ and (b) the initial activity of the sample.   (c) How long does the sample have to be stored if a "safe" activity level is 0.1 Bq?

Solution:

(a)   # nuclei $= \dfrac{\text{mass parent}}{\text{mass of 1 nucleus}}$

$\quad = \dfrac{1 \text{ kg}}{(239 \text{ u})(1.66 \times 10^{-27} \text{ kg/ u})} = 2.52 \times 10^{24}$ □

(b) $l = \dfrac{0.693}{T_{1/2}} = \dfrac{0.693}{(2.4 \times 10^4 \text{ y})(3.156 \times 10^7 \text{ s/ y})}$

$\quad = 9.149 \times 10^{-13} \text{ s}^{-1}$ □

$R_0 = lN_0 = (9.149 \times 10^{-13} \text{ s}^{-1})(2.52 \times 10^{24})$

$\quad = 2.306 \times 10^{12} \text{ decays/s}$ □ $= 2.306 \times 10^{12}$ Bq

(c) $R = R_0 e^{-\lambda t}$ or $e^{-\lambda t} = \dfrac{R}{R_0} = \dfrac{0.1 \text{ Bq}}{2.306 \times 10^{12} \text{ Bq}} = 4.336 \times 10^{-14}$

$\lambda t = -\ln (4.336 \times 10^{-14}) = 30.77$ and

$t = \dfrac{30.77}{\lambda} = \dfrac{30.77}{9.149 \times 10^{-13} \text{ s}^{-1}} = 3.363 \times 10^{13} \text{ s} = 1.07 \times 10^6$ y

$t = 1.07$ million years □

31.59 **Problem:** A large nuclear power reactor produces about 3000 MW of thermal power in its core. Three months after a reactor is shut down, the thermal power in the core is 10 MW, due to radioactive by-products. Assuming that each emission delivers 1 MeV of energy to the thermal power, estimate the activity, in becquerels, three months after the reactor is shut down.

Solution: $P = 10$ MW $= 10^7$ J/s. If each decay delivers 1 MeV $= 1.6 \times 10^{-13}$ J, then the number of decays/s $= (10^7 \text{ J/s})/(1.6 \times 10^{-13} \text{ J/decay}) = 6.25 \times 10^{19}$ Bq $\square$

31.63 **Problems:** Carbon detonations are powerful nuclear reactions that temporarily tear apart the cores of massive stars late in their lives. These blasts are produced by carbon fusion, which requires a temperature of about 6×10^8 K to overcome the strong Coulomb repulsion between carbon nuclei. (a) Estimate the repulsive energy barrier to fusion, using the required ignition temperature for carbon fusion. (In other words, what is the kinetic energy for a carbon nucleus at a temperature of 6×10^8 K?) (b) Calculate the energy (in MeV units) released in each of these "carbon-burning" reactions:

$$^{12}C + {}^{12}C \rightarrow {}^{20}Ne + {}^{4}He$$
$$^{12}C + {}^{12}C \rightarrow {}^{24}Mg + \gamma$$

(c) Calculate the energy (in kilowatt-hours) given off when 2 kg of carbon completely fuses according to the first reaction.

Solutions:

(a) At 6×10^8 K, each carbon nuclei has thermal energy of

$\frac{3}{2} k_B T = (1.5)(8.62 \times 10^{-5} \text{ eV/K})(6 \times 10^8 \text{ K}) = 7.7 \times 10^4$ eV $\square$

(b) Energy released $= [2m(C^{12}) - m(Ne) - m(He^4)]c^2$

$= (24.000000 - 19.992440 - 4.002603)(931.49)$ MeV

$= 4.62$ MeV $\square$

(b) Energy released $= [2m(C^{12}) - m(Mg^{24})](931.49)$ MeV/u

$= (24.000000 - 23.985042)(931.49)$ MeV $= 13.9$ MeV $\square$

(c) Energy released = the energy of reaction of the # of carbon nuclei in a 2 kg sample, which corresponds to

$[(2 \times 10^3$ g $\times 6.02 \times 10^{23}$ atoms/mol/12 g/mol)

(1 fusion event/2 nuclei)(4.62 MeV)1 k $\cdot$ Wh/

(2.25×10^{19} MeV)]

$\Delta E = \dfrac{(1.0 \times 10^{26})(4.62)}{2(2.25 \times 10^{19})}$ kW $\cdot$ h $= 10.6 \times 10^6$ kW $\cdot$ h $\square$

CHAPTER 32

QUESTIONS

2   **Question:** Identify the particle decays in Table 32.2 that occur by the weak interaction. Justify your answer.

Answer: The decays of the muon, tau, positive pion, kaons, neutron, lambda, positive sigma, negative sigma, xis, and omega occur by the weak interaction. All of these have lifetimes of 10^{-13} s or more. Note that several decays result in neutrinos, which do not feel the strong, electromagnetic, or (so far as we know) gravitational forces, but only the weak force. Note that several of these decays involve changes in strangeness, which is conserved by all forces except the weak interaction.

3   **Question:** Identify the particle decays in Table 32.2 that occur by the electromagnetic interaction. Justify your answers.

Answer: The decays of the neutral pion, eta, and neutral sigma occur by the electromagnetic interaction. These are the three shortest lifetimes in Table 32.2. All produce photons, which are the quanta of the electromagnetic force. All conserve strangeness.

9   **Question:** When an electron and a positron meet at low speeds in empty space, why is it that *two* gamma rays with energy of 0.511 MeV are produced, rather than *one* gamma ray with energy of 1.02 MeV?

Answer: Photons carry momentum. If only one were produced, momentum could not be conserved. The two actually produced travel off in opposite directions.

10   **Question:** Why is it that the neutron (which decays in free space in 900 s) is stable inside the nucleus?

Answer: A neutron inside a nucleus is stable because it is in a lower-energy state than a free neutron, and lower in energy than it would be if it decayed into a proton (plus electron and antineutrino). The nuclear force gives it this lower energy by binding it inside the nucleus, and by favoring pairing between neutrons and protons. Saying that the zone of stability is narrow in a neutron-proton plot of stable nuclei (Figure 31.3) is another way of saying that a neutron cannot decay to a proton in a stable nucleus.

12   **Question:** What is the quark composition of the Ξ^- particle? (See Table 32.4.)

Answer: The xi-minus particle has, from Table 32.2, charge $-e$, spin $\hbar/2$, $B = 1$, $L_e = L_\mu = L_\tau = 0$, and strangeness -2. All these are described by its quark composition dss (Table 32.4). The properties of the quarks from Table 32.3 let us add up charge: $-e/3 - e/3 - e/3 = -e$; spin $+ \hbar/2 - \hbar/2 + \hbar/2 = \hbar/2$ (supposing one of the quarks is spin-down relative to the other two); baryon number $\frac{1}{3} + \frac{1}{3} + \frac{1}{3} = 1$; lepton numbers, charm, bottomness, topness zero; and strangeness $0 - 1 - 1 = -2$.

CHAPTER 32

PROBLEMS

32.3 **Problem:** A photon with an energy of E_γ = 2.09 GeV creates a proton-antiproton pair in which the proton has a kinetic energy of 95 MeV. What is the kinetic energy of the antiproton? ($m_p c^2$ = 938.3 MeV.)

Solution: In $\gamma \to p^+ + p^-$, we start with energy 2.09 GeV, we end with energy 938.3 MeV + 938.3 MeV + 95 MeV + K_2 where K_2 is the kinetic energy of the second proton. Conservation of energy gives K_2 = 118.4 MeV $\square$

32.7 **Problem:** A neutral pi meson at rest decays into two gamma-ray photons according to

$$\pi^0 \to \gamma + \gamma$$

Find the energy, momentum, and frequency of each gamma-ray photon.

Solution: Since the pion is at rest and momentum is conserved, the two gamma-rays must have equal momenta in opposite directions. So they must share equally in the energy of the pion.

M_{π^0} = 135 MeV/c^2 (Table 32.2)

$\therefore E_\gamma$ = 67.5 MeV $\square$

$$p = \frac{E}{c} = 67.5 \text{ MeV/c } \square; \quad f = \frac{E}{h} = 1.63 \times 10^{22} \text{ Hz } \square$$

32.9 Problem: Name one possible decay mode (see Table 32.2) of each of the following particles: Ω^+, $\overline{K^0}$, $\overline{\Lambda^0}$, $\bar{n}$

Solution: These decay into the antiparticles of the listed particles into which their antiparticles decay:

$$\Omega^+ \to \overline{\Lambda^0} + K^+$$

$$\overline{K^0} \to \pi^+ + \pi^- \text{ or } \pi^0 + \pi^0$$

$$\overline{\Lambda^0} \to \bar{p} + \pi^+$$

$$\bar{n} \to \bar{p} + e^+ + \nu_e$$

32.11 Problem: Each of the following reaction is forbidden. Determine a conservation law that is violated by each reaction.

(a) $p^+ + p^- \to \mu^+ + e^-$

(b) $\pi^- + p^+ \to p^+ + \pi^+$

(c) $p^+ + p^+ \to p^+ + \pi^+$

(d) $p^+ + p^+ \to p^+ + p^+ + n$

(e) $\gamma + p^+ \to n + \pi^0$

Solution: (a) $p + \bar{p} \to m^+ + e^-$ violates conservation of the lepton numbers L_e: $0 + 0 \to 0 + 1$ and L_μ: $0 + 0 \to -1 + 0$

(b) $\pi^- + p \to p + \pi^+$ charge: $-1 + 1 \to +1 + 1$

(c) $p + p \to p + \pi^+$ baryon number: $1 + 1 \to 1 + 0$

(d) $p + p \to p + p + n$ baryon number: $1 + 1 \to 1 + 1 + 1$

(e) $\gamma + p \to n + \pi^0$ charge: $0 + 1 \to 0 + 0$

32.17 **Problem:** Determine whether or not strangeness is conserved in the following decays and reactions.

(a) $\Lambda^0 \rightarrow p^+ + \pi^-$

(b) $\pi^- + p^+ \rightarrow \Lambda^0 + K^0$

(c) $p^- + p^+ \rightarrow \overline{\Lambda^0} + \Lambda^0$

(d) $\pi^- + p^+ \rightarrow \pi^- + \Sigma^+$

(e) $\Xi^- \rightarrow \Lambda^0 + \pi^-$

(f) $\Xi^0 \rightarrow p^+ + \pi^-$

Solution: We look up the strangeness quantum numbers in Table 32.2.

(a) $\Lambda^0 \rightarrow p + \pi^-$

Strangeness: $-1 \rightarrow 0 + 0$ (strangeness is not conserved)

(b) $\pi^- + p \rightarrow \Lambda^0 + K^0$

Strangeness: $0 + 0 \rightarrow -1 + 1$ ($0 = 0$ and strangeness is conserved)

(c) $\overline{p} + p \rightarrow \overline{\Lambda^0} + \Lambda^0$

Strangeness: $0 + 0 \rightarrow +1 - 1$ ($0 = 0$ and strangeness is conserved)

(d) $\pi^- + p \rightarrow \pi^- + \Sigma^+$

Strangeness: $0 + 0 \rightarrow 0 - 1$ (0 does not equal -1 so strangeness is not conserved)

(e) $\Xi^- \rightarrow \Lambda^0 + \pi^-$

Strangeness: $-2 \rightarrow -1 + 0$ (-2 does not equal -1 so strangeness is not conserved)

(f) $\Xi^0 \rightarrow p + \pi^-$

Strangeness: $-2 \rightarrow 0 + 0$ (-2 does not equal 0 so strangeness is not conserved)

32.23 Problem: Analyze each of the reactions in terms of its constituent quarks:

(a) $\pi^- + p^+ \rightarrow K^0 + \Lambda^0$ (c) $K^- + p^+ \rightarrow K^+ + K^0 + \Omega^-$

(b) $\pi^+ + p^+ \rightarrow K^+ + \Sigma^+$ (d) $p^+ + p^+ \rightarrow K^0 + p^+ + \pi^+ + ?$

In the last reaction, identify the mystery particle.

Solution: We look up the quark constituents of the particles in Table 32.4:

(a) $d\bar{u} + uud \rightarrow d\bar{s} + uds$

(b) $d\bar{u} + uud \rightarrow u\bar{s} + uus$

(c) $u\bar{s} + uud \rightarrow u\bar{s} + d\bar{s} + sss$

(d) $uud + uud \rightarrow d\bar{s} + uud + u\bar{d} + uds$ □

    A uds is a Λ^0

32.31 Problem: What are the kinetic energies of the proton and pi meson resulting from the decay of a Λ^0 at rest?
$\Lambda^0 \rightarrow p^+ + \pi^-$

Solution:

$m_\Lambda c^2 = 1115.6$ MeV $\Lambda^0 \rightarrow p + \pi^-$

$m_p c^2 = 938.3$ Mev

$m_\pi c^2 = 139.6$ Mev

The difference between starting mass-energy and final mass-energy is the kinetic energy of the products.

$(K_p + K_\pi) = 37.7$ MeV and momentum is conserved: $|p_p| = |p_\pi| = p$
Applying conservation of relativistic energy.

$$\left(\sqrt{(938.3)^2 + p^2 c^2} - 938.3\right) + \sqrt{(139.6)^2 + p^2 c^2} - 139.6 = 37.7 \text{ MeV}$$

Solving the algebra yields $p_\pi c = p_p c = 100.4$ MeV

Then

$$K_p = \sqrt{(m_p c^2)^2 + (100.4)^2} - m_p c^2 = 5.4 \text{ MeV } \square$$

$$K_\pi = \sqrt{(139.6)^2 + (100.4)^2} - 139.6 = 32.3 \text{ MeV } \square$$

32.33 **Problem:** If a K^0 meson at rest decays in 0.9×10^{-10} s, how far will a K^0 meson travel if it is moving at $0.96c$ through a bubble chamber?

Solution: The K^0 is relativistic, with time-dilated lifetime

$$T = \gamma T_0 = \frac{0.9 \times 10^{-10} \text{ s}}{\sqrt{1 - v^2 / c^2}} = \frac{0.9 \times 10^{-10} \text{ s}}{\sqrt{1 - (0.96)^2}} = 3.214 \times 10^{-10} \text{ s}$$

distance $= (0.96)(3 \times 10^8 \text{ m/s})(3.214 \times 10^{-10} \text{ s}) = 9.26 \text{ cm} \; \square$